Discovering Advanced Algebra

An Investigative Approach

SECOND EDITION

Teaching and Worksheet Masters

DISCOVERING MATHEMATICS

Editor: Elizabeth DeCarli

Project Administrator: Tamar Wolins

Production Editor: Christa Edwards

Editorial Production Supervisor: Kristin Ferraioli

Production Director: Christine Osborne

Senior Production Coordinator: Ann Rothenbuhler

Text Designer: Jenny Somerville

Composition, Technical Art: ICC Macmillan Inc.

Cover Designer: Jill Kongabel, Jeff Williams

Printer: Versa Press, Inc.

Textbook Product Manager: Tim Pope

Executive Editor: Josephine Noah

Publisher: Steven Rasmussen

Cover Photo Credits: Background and center images: NASA; all other images: Ken Karp Photography.

Key Curriculum Press
1150 65th Street
Emeryville, CA 94608
(510) 595-7000
editorial@keypress.com
www.keypress.com

Printed in the United States of America
10 9 8 7 6 5 4 3 2 1 13 12 11 10 09

ISBN 978-1-60440-007-6

Contents

Chapter 4

Chapter 5

Chapter 6

Chapter 7

Chapter 8

Chapter 9

Chapter 10

Chapter 11

Chapter 12

Chapter 13

Materials

Introduction

These masters for *Discovering Advanced Algebra: An Investigative Approach* are designed for your students to use as they work on investigations or exercises, or for you to use in your classroom presentations and discussions. Many of the one-step investigations have transparency or worksheet masters so you can provide a visual aid to students as they work through the problems you give them. Sample data transparencies are useful when you don't have the time or equipment needed to gather data in class. The *Discovering Advanced Algebra Teacher's Edition* lists lessons that include masters in the introduction before the beginning of each chapter, and names the masters in the materials list at the beginning of each lesson.

Transparency masters contain graphs and diagrams from the student book, enlarged so that they can be seen easily when displayed on an overhead projector. Worksheet masters are designed to help students perform investigations or work on special projects. Worksheets often contain large type so that you can also use them as transparencies. If a worksheet is required for a particular investigation, it is listed in the student book and in the materials list in the margin of the *Teacher's Edition.* Optional worksheets and transparencies are listed only in the *Teacher's Edition* and are followed by the word *optional.*

The masters are placed in the first lesson in which they appear. Some, such as the Algebra Tiles worksheet, may be used with many different lessons. Masters for several different sizes of graph paper, as well as masters for rulers and protractors, are also included. Individual worksheets for every investigation are provided in a separate resource, *Investigation Worksheets.* Each worksheet includes all of the steps and questions from the investigation, with space for student work and, where appropriate, blank tables and graphing grids.

Coordinate Axes

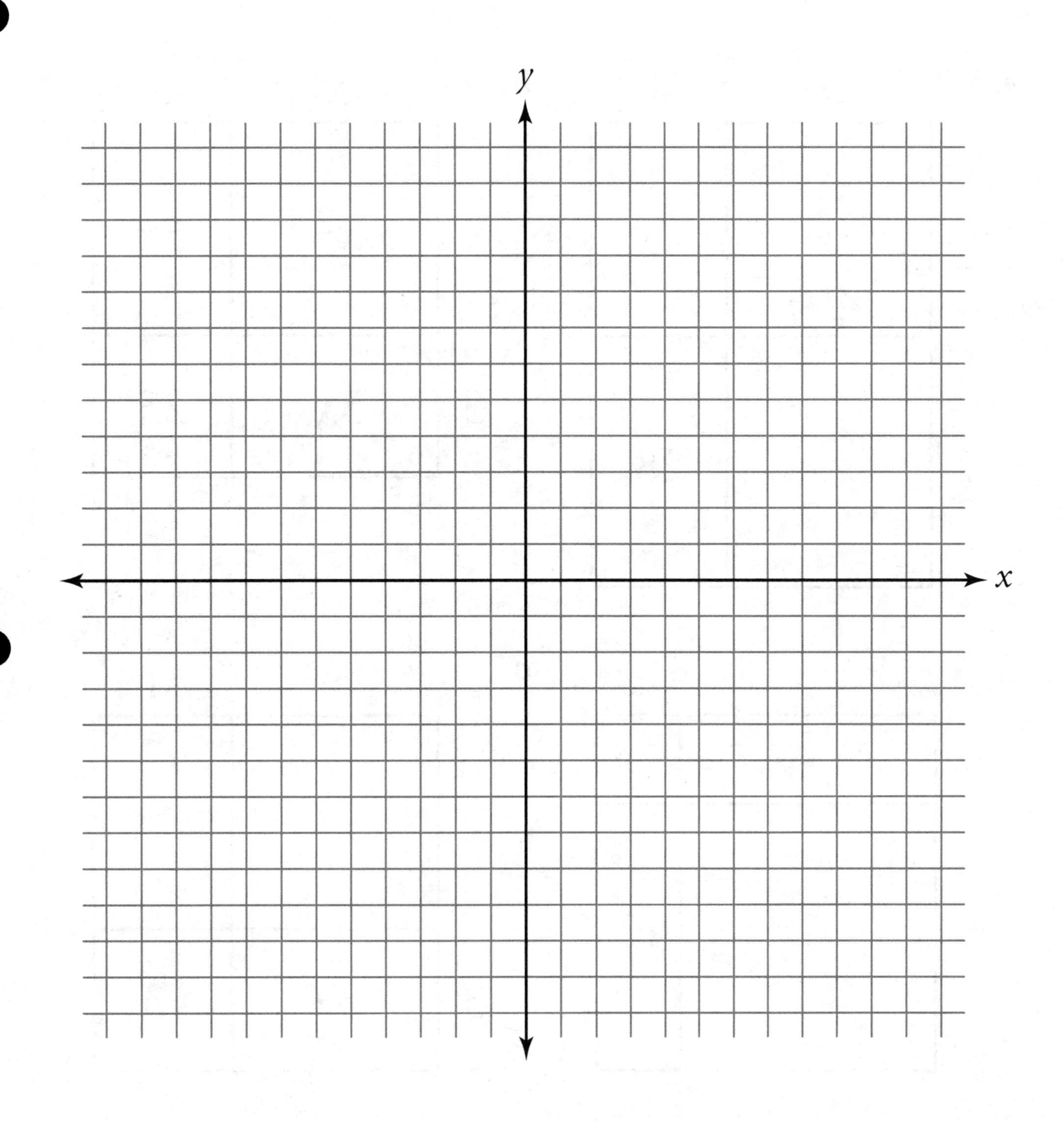

Exercise 11

a.

	x	4
x	x^2	$4x$
7	$7x$	28

b.

	x	5
x	x^2	$5x$
5	$5x$	25

c.

	x	2
y	xy	$2y$
6	$6x$	12

d.

	x	-1
x	x^2	$-1x$
3	$3x$	-3

Algebra Tiles

Exercise 9

Name ______________________ Period __________ Date ______________

	Erlenmeyer flask	Round flask	Bottle	Calcium	Sulfur	Iron	Lake	Well	Spring
Left									
Center									
Right									
Lake									
Well									
Spring									
Calcium									
Sulfur									
Iron									

Sample of a Lab Report

Include the names of all group members, with your name listed first. In Section I, describe what you studied in the lab, including both the procedure and the mathematics that you used. In Section II, record the complete set of data. (You can use a computer to print out long data sets from your calculator.) In Section III, provide any graphs of your data that you have made. In Sections IV and V, show your work for all calculations and your conclusions. If you run across anything unusual, make a note of it.

Name Your name, Group names **Period** ______ **Date** ______

I. Overview

We lined up desks and experimented with how many more people could fit each time a desk was added. We started with 1 desk, which seated 4 people, and ended with 10 desks, which seated 22 people. We made a data table to summarize our observations, drew a graph, and wrote a recursive formula that models the data in the table.

II. Data Table

Number of desks	1	2	3	4	5	6	7	8	9	10
Number of people	4	6	8	10	12	14	16	18	20	22

III. Graph(s)

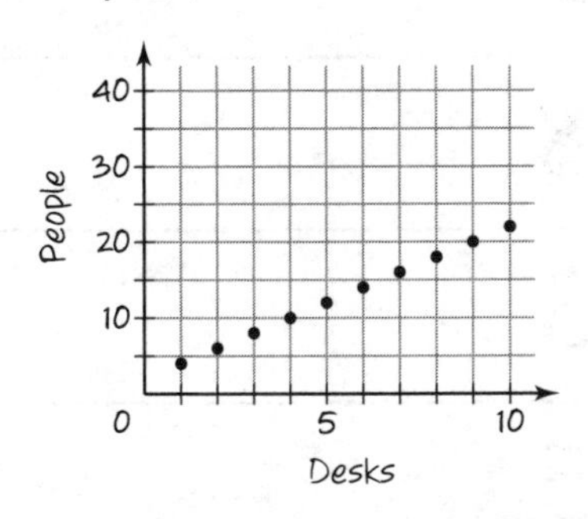

IV. Equations, calculations, and formulas

$u_1 = 4$, because 4 people can fit at 1 desk

$u_n = u_{n-1} + 2$, because 2 people are always added to the previous term

$n \geq 2$, because the recursive rule is first used to find u_2

recursive formula: $u_1 = 4$, $u_n = u_{n-1} + 2$ where $n \geq 2$

V. Conclusions

The graph shows a clear pattern. The line can be extended to find the number of desks needed for a given number of people. This is a recursive process; two people are repeatedly added to the previous number of people. The number of people at 1 desk is 4. The number of people at n desks is the number of people at (n - 1) desks plus 2.

Sample Market Data

Time period	Number of people who own the gadget
0	1
1	2
2	4
3	7
4	12
5	18
6	23
7	25

Backpacks

Student	Grade	Weight of backpack (lb)
1	Junior	10
2	Senior	20
3	Junior	9
4	Junior	17
5	Junior	3
6	Junior	10
7	Senior	15
8	Junior	15
9	Senior	7
10	Senior	10
11	Junior	9
12	Senior	10
13	Senior	9
14	Junior	7
15	Senior	4

Student	Grade	Weight of backpack (lb)
16	Senior	6
17	Senior	7
18	Senior	9
19	Junior	13
20	Junior	10
21	Senior	8
22	Senior	7
23	Senior	4
24	Senior	4
25	Junior	8
26	Junior	33
27	Senior	10
28	Senior	9
29	Senior	7
30	Junior	16

Box Plot

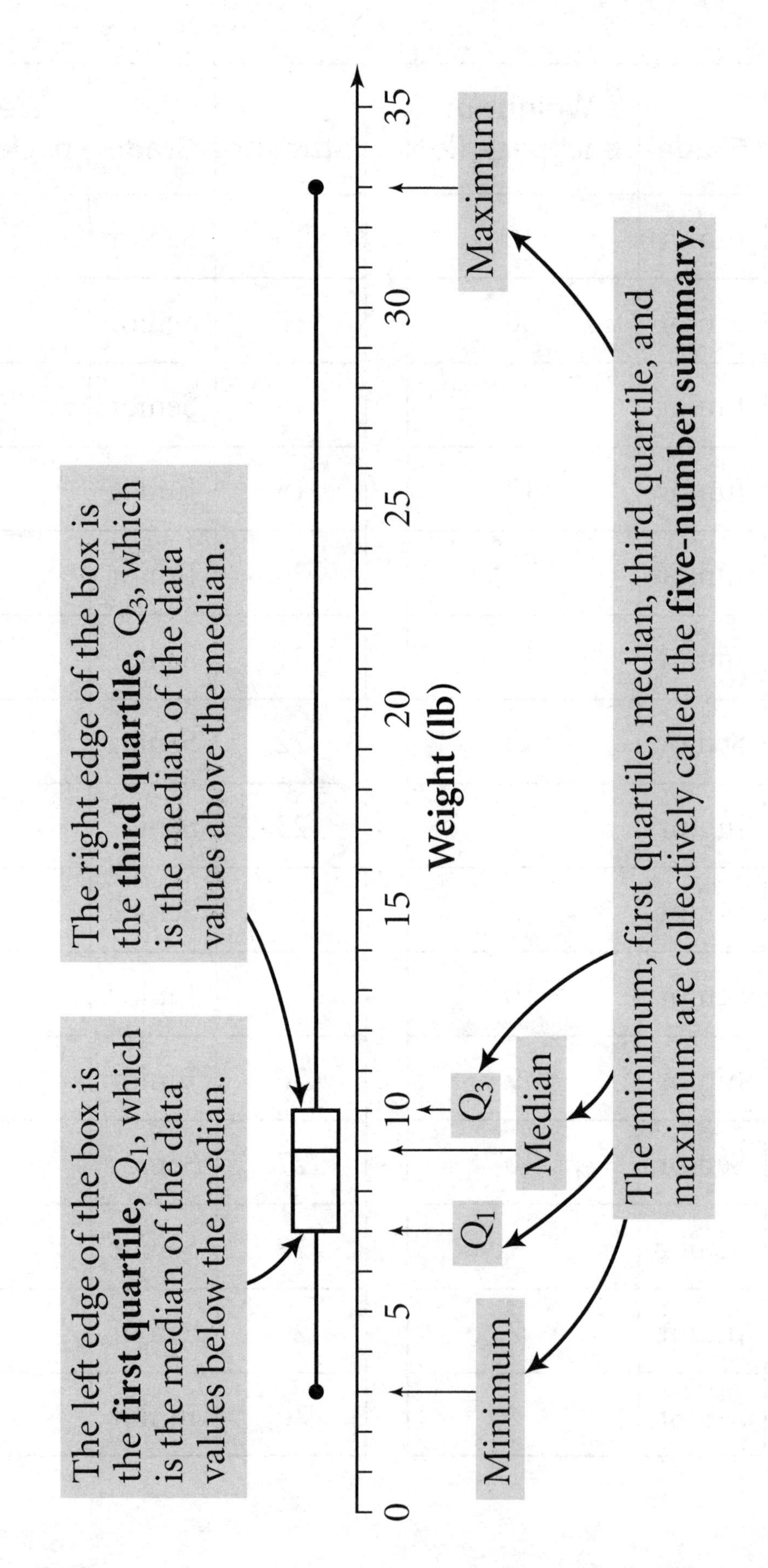
The left edge of the box is the first quartile, Q_1, which is the median of the data values below the median.
The right edge of the box is the third quartile, Q_3, which is the median of the data values above the median.
The minimum, first quartile, median, third quartile, and maximum are collectively called the five-number summary.
Minimum
Q_1
Median
Q_3
Maximum
0
5
10
15
20
25
30
35
Weight (lb)

Modified Box Plot

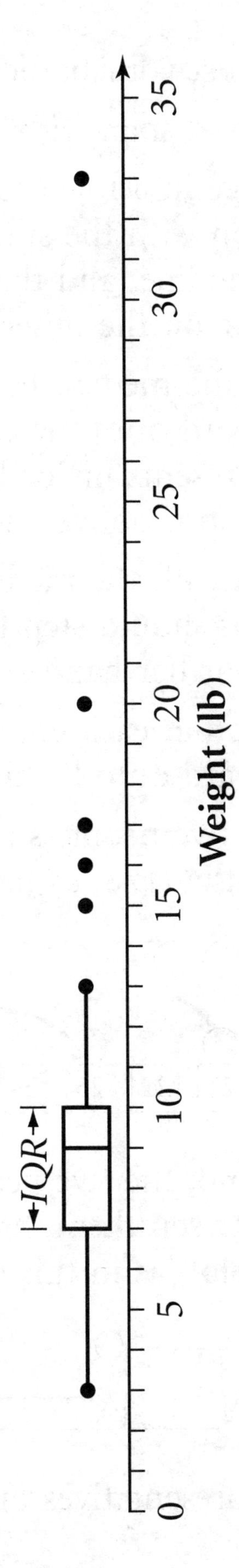

Human Box Plot

Create a box plot out of just your bodies and some rope.

Step 1 Measure your hand span with a centimeter ruler.

Step 2 As a class, arrange in a line according to your hand spans. The person with the smallest hand span will be on one end of the line, and the person with the largest hand span will be on the other end.

Step 3 Decide who has the median hand span. This person should step forward onto the card labeled with the number that represents his or her hand span. There are now two groups, one above and one below the median.

Next, the person with the median hand span in each of the two groups should step forward onto the card representing his or her hand span.

Finally, the person at each end of the original line should step forward onto his or her card.

There should be five people standing in a line apart from the rest of the class, as in this diagram.

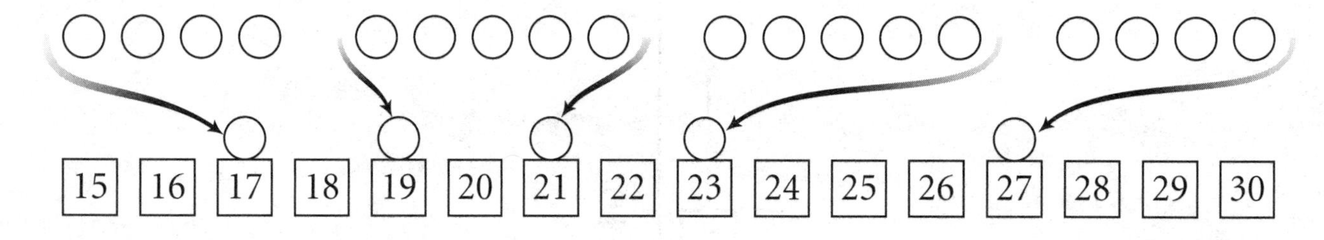

Step 4 Starting at one end, the five people on the cards should pass the rope between them, holding the rope to resemble a box plot, as in this diagram.

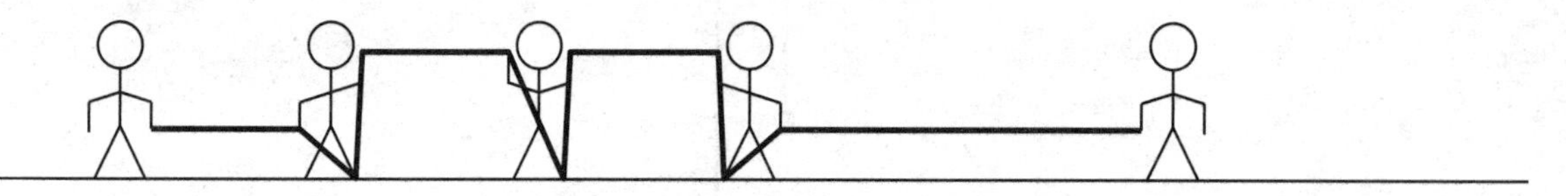

These five students are representatives of the whole group.

Spread

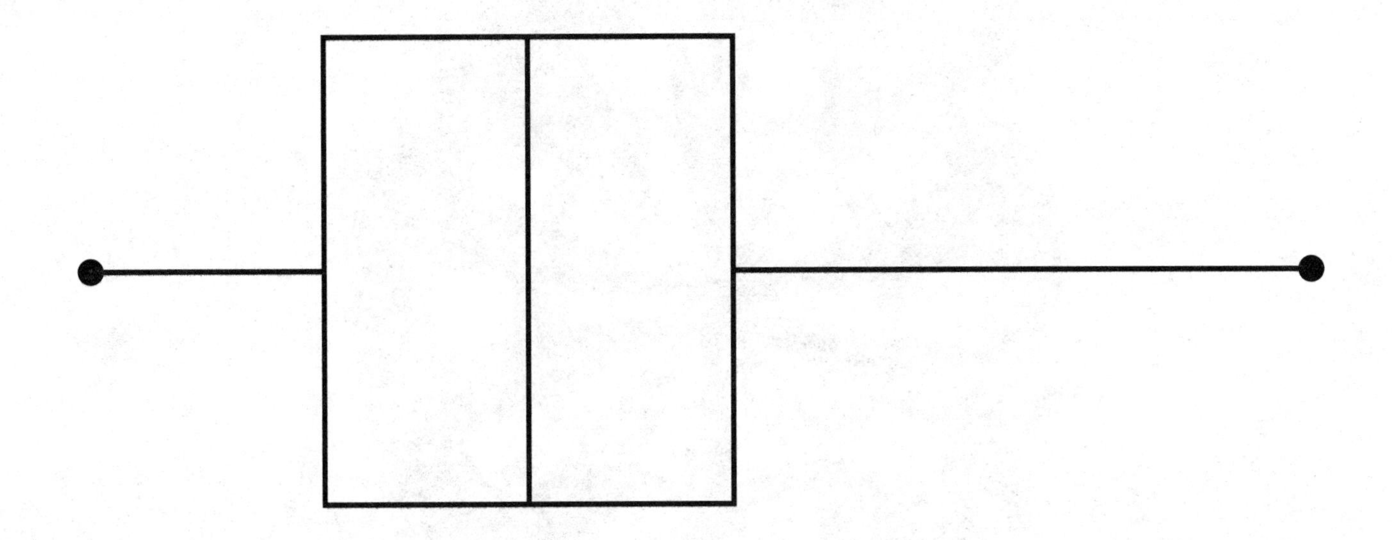

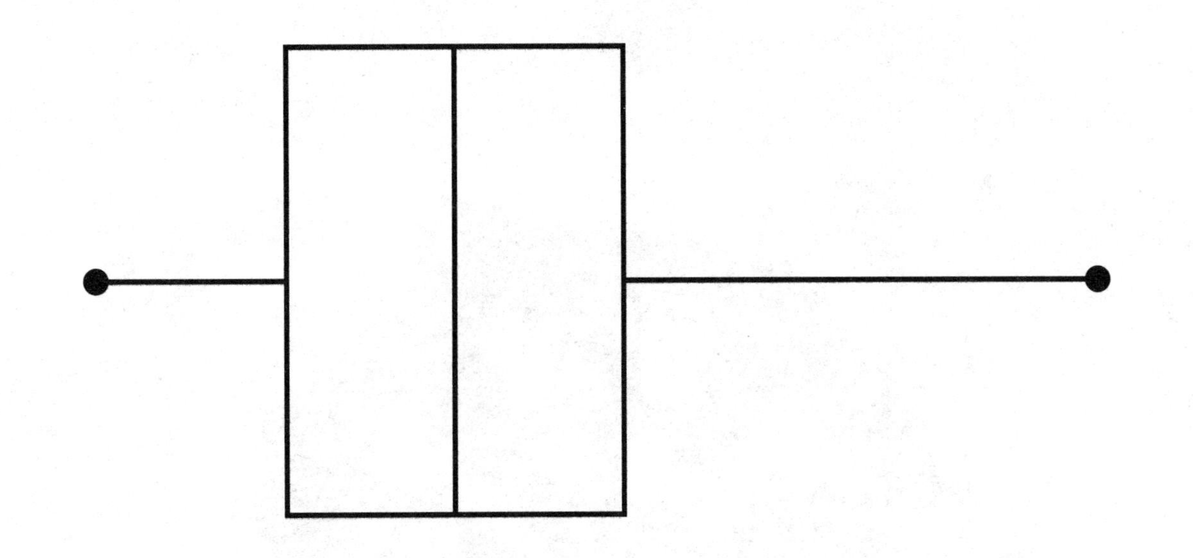

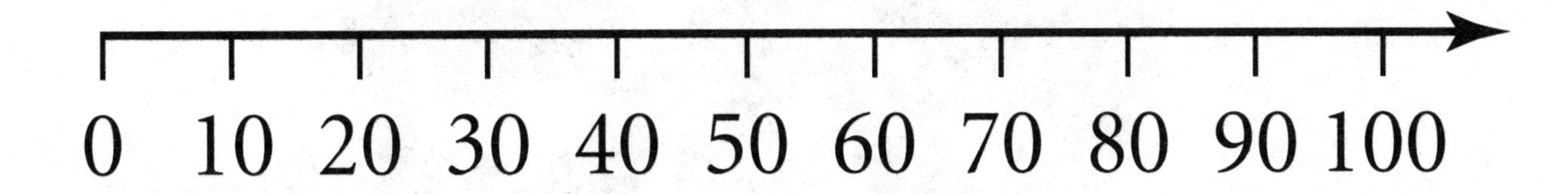

Dartboards

17
6
10
18
2
1
13
20
3
5
19
12
7
16
14
8
17
6
4
15
18
2
13
3
5
19
12
7
9
16
14
11
8

Class Measurements of Textbook

	Length (cm)	Width (cm)	Depth (cm)	Calculated volume (cu. cm)
1				
2				
3				
4				
5				
6				
7				
8				
9				
10				
11				
12				
13				
14				
15				
16				
17				
18				
19				
20				

Experiments 1 and 2 Sample Data

Rolling Ball Sample Data

Distance (cm)	17.3	24.4	28.7	19.2	36.8	30.8	41.6

Rubber Band Sample Data

Distance (cm)	182.2	135.9	187.6	162.5	150.0	186.5	180.0

Student-to-Teacher Ratios Dot Plot

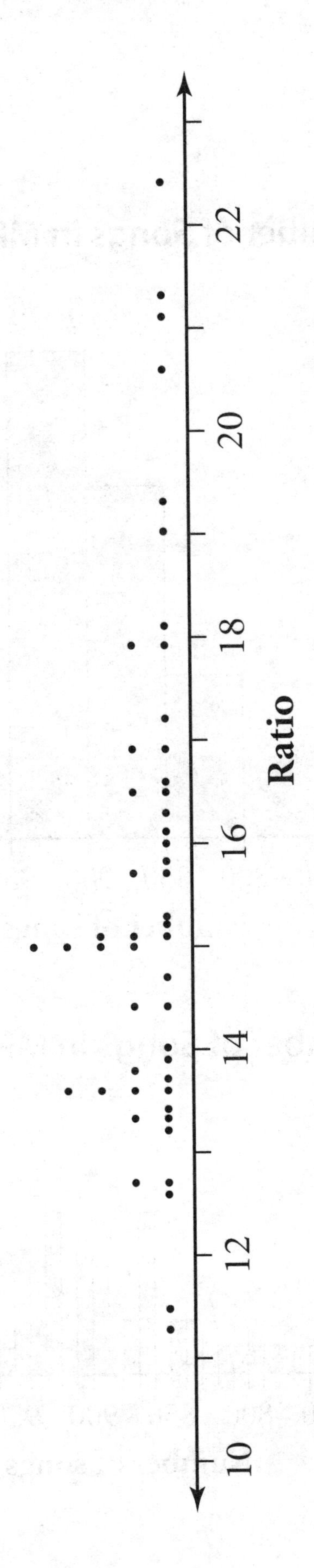

Songs in MP3 Player

Graph A

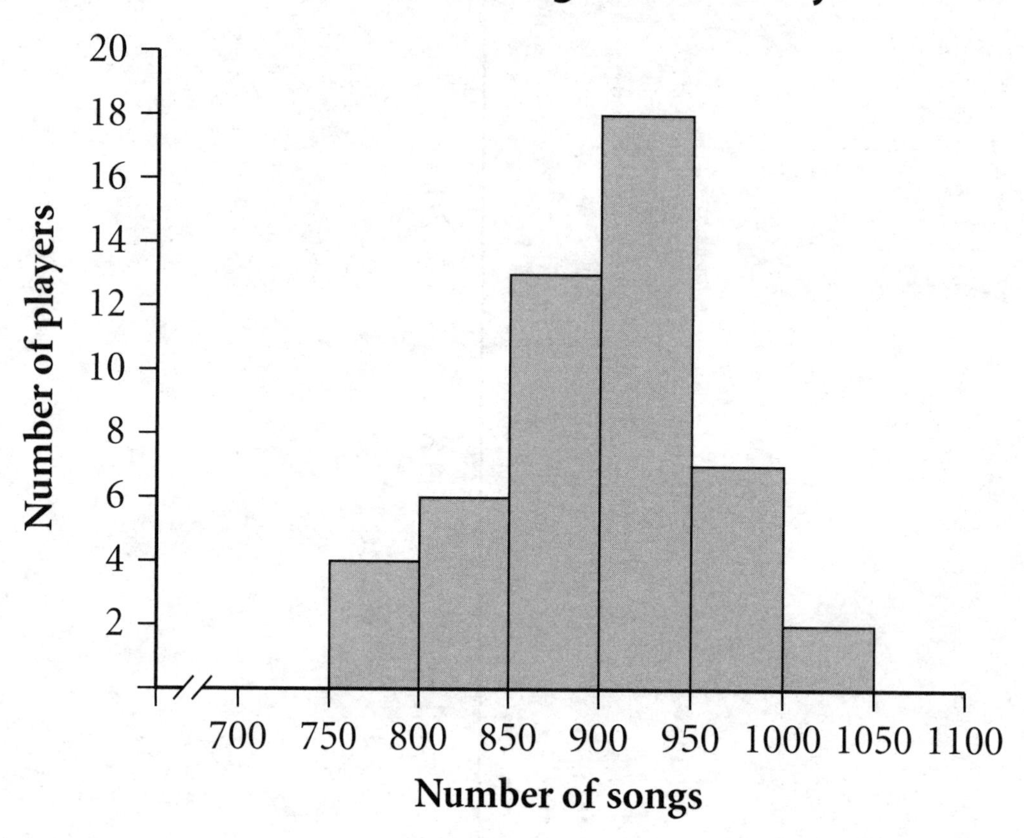

Graph B

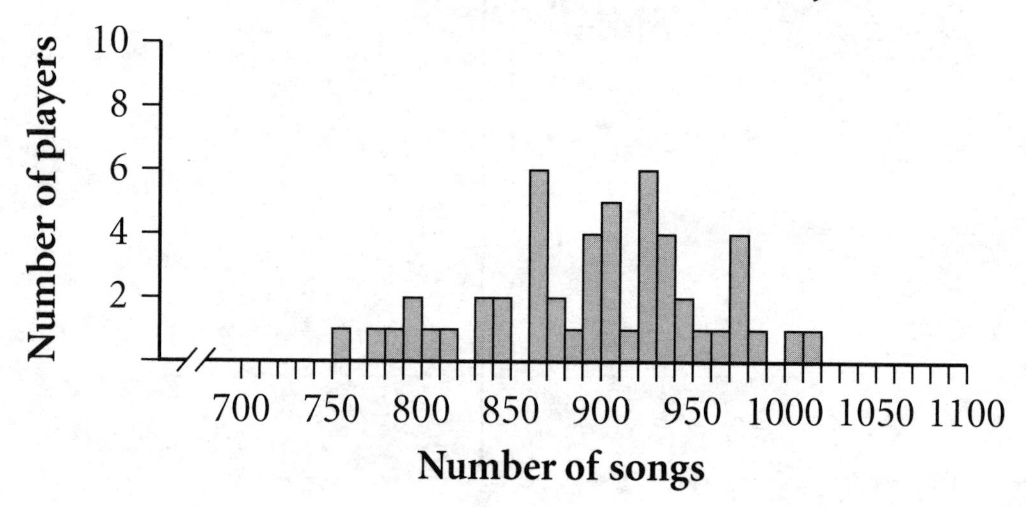

Percentile Rank

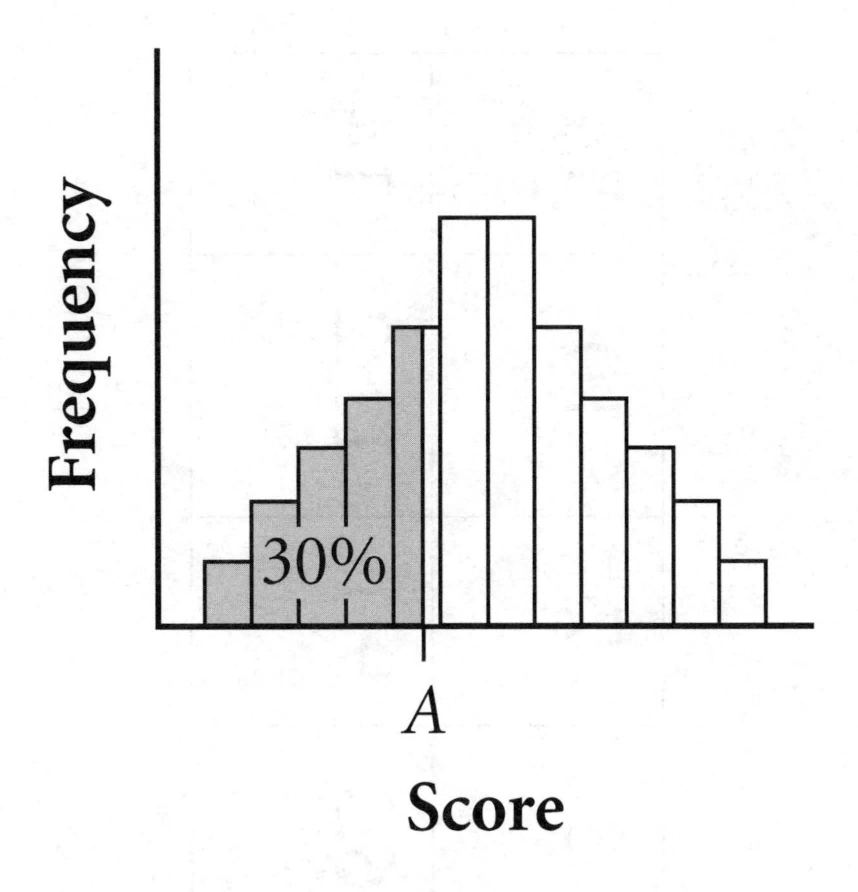

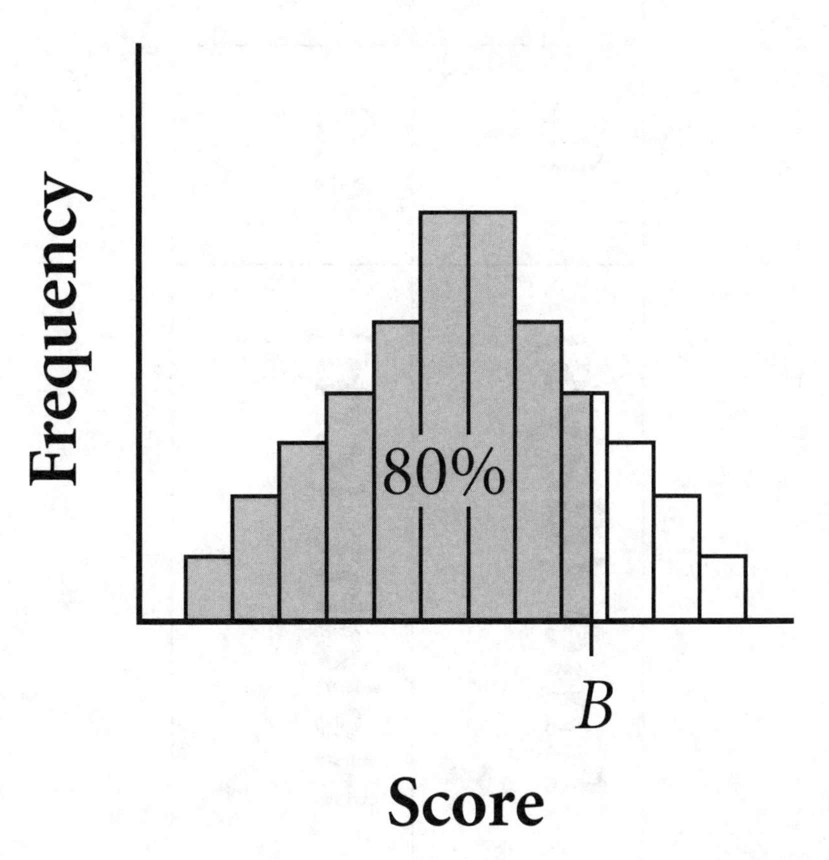

Growing a Lawn

Time (days) x	0	3	7	10	14
Height (cm) y	4.2	6.3	9.1	11.2	14

Wave Sample Data

Number of people x	2	5	6	8	9	10	15	16	22
Time (s) y	2.1	4.4	5.2	5.8	4.7	6.7	7.5	10.4	11.0

Exercise 7

a.

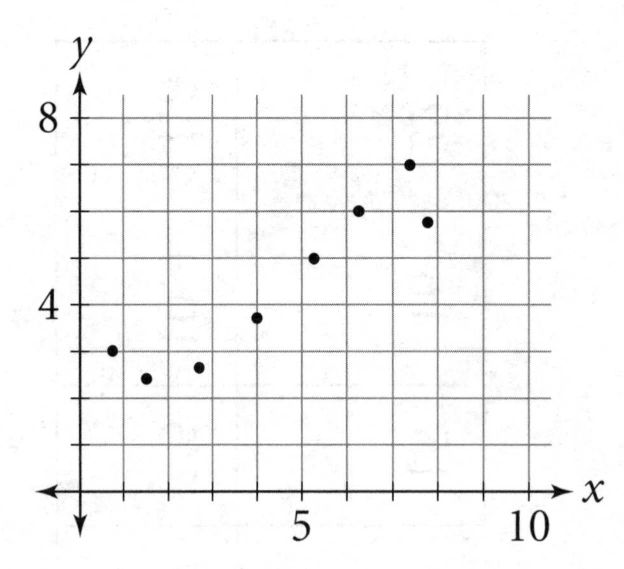

b.

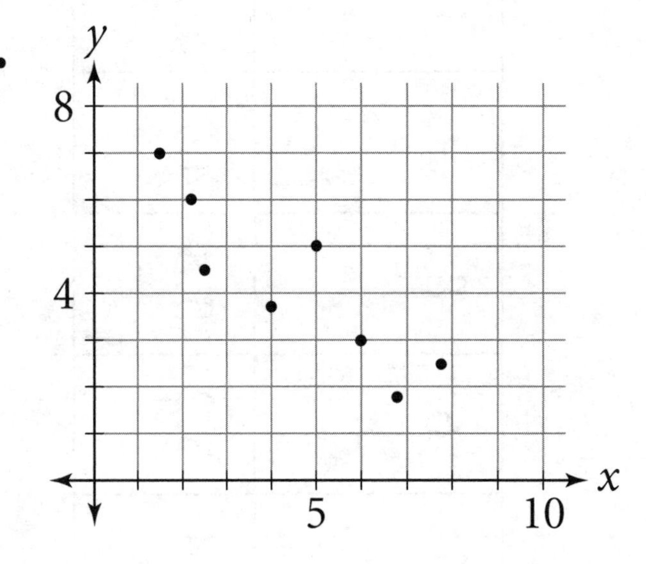

c.

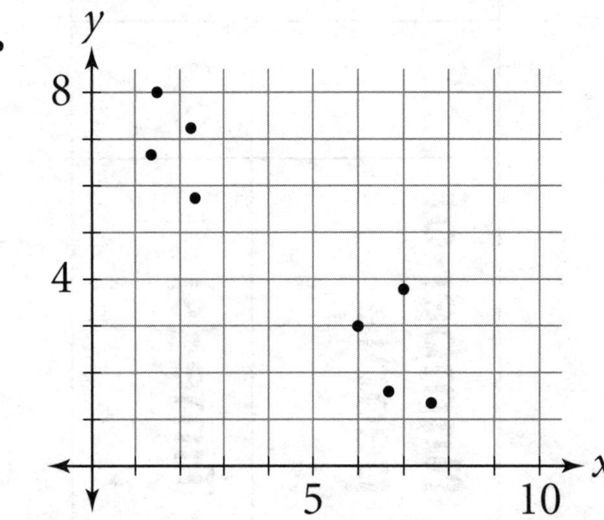

Year	Population (millions)	Waste (million tons/yr)
1960	179	88.1
1970	203	121.1
1980	227	151.6
1990	249	205.2
2000	281	238.3
2006	300	251.3

(*www.epa.gov*)

Median-Median Line

x	y
4.0	23
5.2	28
5.8	29
6.5	35
7.1	35
7.8	40
8.3	42

x	y
8.9	50
9.7	47
10.4	52
11.2	60
11.9	58
12.5	62
13.1	64

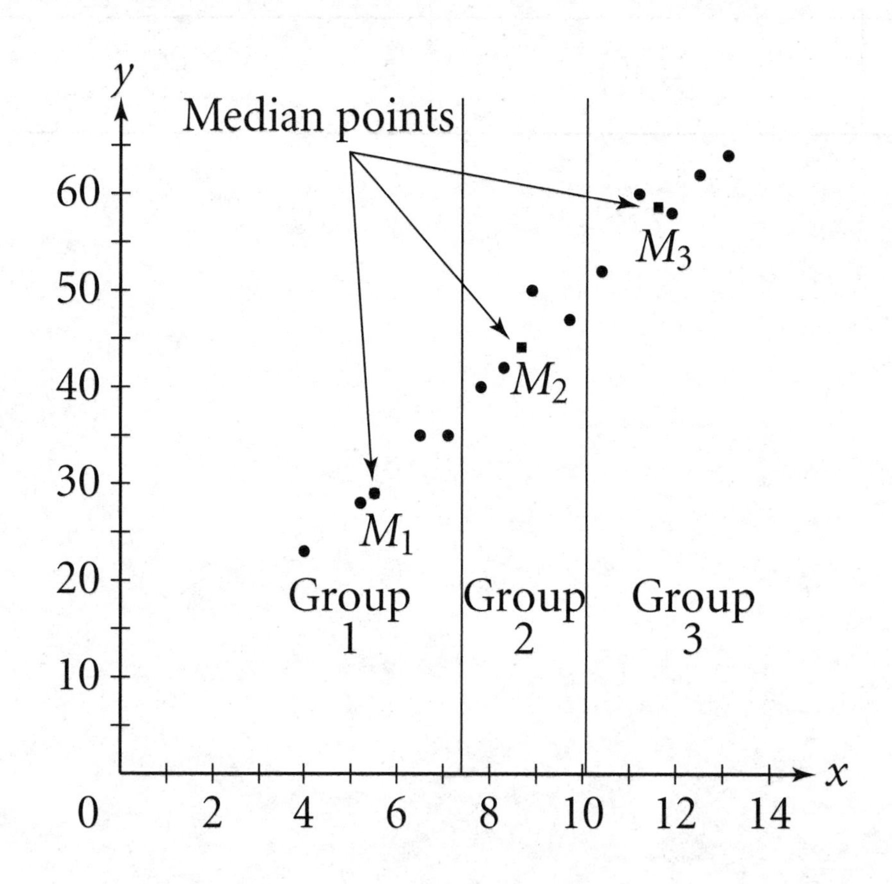

Airline Schedules (page 1 of 2)

DEP—Departure time from origination city shown in that city's local time
720a = 7:20 A.M. (a = A.M., p = P.M.)

ARR—Arrival time in destination city shown in that city's local time

MI—Approximate mileage

Airport Codes

City	Code
Columbus, OH	CMH
Charleston, WV	CRW
Cincinnati, OH	CVG
Wausau/Stevens Point, WI	CWA
Washington, DC (Nat'l)	DCA
Denver, CO	DEN
Dallas/Fort Worth, TX	DFW
Des Moines, IA	DSM
Detroit, MI	DTW
Escanaba, MI	ESC
New York (Newark), NJ	EWR
Fort Wayne, IN	FWA
Grand Forks, ND	GFK
Greenville/Spartanburg, SC	GSP
Houston (Hobby), TX	HOU
Indianapolis, IN	IND
Jacksonville, FL	JAX
Lansing, MI	LAN
Las Vegas, NV	LAS
Los Angeles, CA	LAX
New York (LaGuardia), NY	LGA
Kansas City, MO	MCI
Orlando, FL	MCO
Harrisburg, PA	MDT
Memphis, TN	MEM

City	Code
Miami, FL	MIA
Milwaukee, WI	MKE
Muskegon, MI	MKG
Madison, WI	MSN
Minneapolis/St. Paul, MN	MSP
New Orleans, LA	MSY
Chicago (O'Hare), IL	ORD
Norfolk, VA	ORF
West Palm Beach, FL	PBI
Portland, OR	PDX
Philadelphia, PA	PHL
Phoenix, AZ	PHX
Pittsburgh, PA	PIT
Providence/Newport, RI	PVD
Raleigh/Durham, NC	RDU
Richmond, VA	RIC
Roanoke, VA	ROA
Rochester, NY	ROC
San Diego, CA	SAN
Louisville, KY	SDF
Seattle/Tacoma, WA	SEA
San Francisco, CA	SFO
St. Louis, MO	STL
Traverse City, MI	TVC
Knoxville, TN	TYS

Airline Schedules (page 2 of 2)

Flight Schedule

From DTW to:	DEP	ARR	MI
CMH	655p	747p	156
CMH	1000p	1049p	156
CRW	1045a	1215p	281
CRW	435p	605p	281
CRW	845p	1015p	281
CVG	700a	820a	229
CVG	1200p	104p	229
CVG	840p	944p	229
CWA	145p	235p	363
CWA	510p	600p	363
CWA	830p	915p	363
DCA	710a	838a	405
DCA	1210p	131p	405
DCA	835p	1004p	405
DEN	920a	1022a	1125
DEN	1215p	123p	1125
DEN	700p	803p	1125
DFW	700a	848a	986
DFW	1215p	209p	986
DFW	645p	841p	986
DSM	1220p	113p	533
DSM	640p	729p	533
ESC	930a	1115a	305
ESC	1215p	150p	305
EWR	700a	833a	488
EWR	150p	324p	488
EWR	845p	1027p	488
FWA	915a	900a	128
FWA	1235p	1225p	128
FWA	645p	645p	128
FWA	1010p	953p	128
GFK	1235p	329p	782
GSP	715a	851a	508
GSP	135p	308p	508
GSP	825p	958p	508
HOU	930a	1139a	1092
HOU	1230p	236p	1092
HOU	645p	846p	1092
IND	715a	723a	230
IND	1225p	1230p	230
IND	445p	451p	230
IND	1005p	1006p	230
JAX	925a	1141a	814
JAX	1205p	224p	814
JAX	505p	725p	814
LAN	920a	958a	74
LAN	1220p	100p	74
LAN	510p	550p	74
LAN	1000p	1040p	74

From DTW to:	DEP	ARR	MI
LAS	935a	1056a	1749
LAS	1215p	135p	1749
LAS	320p	431p	1749
LAS	1010p	1127p	1749
LAX	925a	1110a	1979
LAX	1220p	217p	1979
LAX	650p	838p	1979
LAX	1005p	1153p	1979
LGA	700a	838a	501
LGA	150p	332p	501
LGA	440p	624p	501
LGA	700p	844p	501
LGA	840p	1026p	501
MCI	725a	822a	630
MCI	925a	1022a	630
MCI	1225p	127p	630
MCI	420p	519p	630
MCI	655p	757p	630
MCO	730a	1009a	957
MCO	900a	1142a	957
MCO	1045a	127p	957
MCO	130p	400p	957
MCO	500p	727p	957
MCO	850p	1125p	957
MDT	1035a	1215p	370
MDT	515p	635p	370
MDT	830p	949p	370
MEM	700a	758a	610
MEM	1200p	1257p	610
MEM	315p	406p	610
MEM	1005p	1049p	610
MIA	720a	1023a	1145
MIA	500p	805p	1145
MKE	940a	950a	237
MKE	310p	315p	237
MKE	705p	712p	237
MKE	1000p	1000p	237
MKG	940a	1045a	161
MKG	100p	205p	161
MKG	845p	950p	161
MSN	935a	951a	312
MSN	1220p	1234p	312
MSN	1000p	1009p	312
MSP	925a	1024a	528
MSP	1020a	1105a	528
MSP	130p	215p	528
MSP	535p	629p	528
MSP	1010p	1056p	528
MSY	940a	1119a	926

From DTW to:	DEP	ARR	MI
MSY	705p	834p	926
ORD	800a	820a	235
ORD	1220p	1239p	235
ORD	1010p	1014p	235
ORF	710a	849a	529
ORF	140p	318p	529
ORF	820p	1002p	529
PBI	920a	1216p	1086
PBI	835p	1128p	1086
PDX	645p	834p	1953
PHL	705a	840a	453
PHL	145p	315p	453
PHL	830p	1007p	453
PHX	930a	1042a	1671
PHX	655p	803p	1671
PIT	715a	814a	201
PIT	315p	413p	201
PIT	820p	922p	201
PVD	705a	847a	614
PVD	450p	633p	614
PVD	820p	1001p	614
RDU	710a	849a	502
RDU	145p	319p	502
RDU	840p	1017p	502
RIC	715a	843a	456
RIC	145p	316p	456
RIC	820p	950p	456
ROA	715a	905a	382
ROA	845p	1035p	382
ROC	840a	946a	296
ROC	820p	928p	296
SAN	930a	1115a	1956
SAN	655p	845p	1956
SDF	440p	550p	306
SDF	1000p	1107p	306
SEA	930a	1127a	1927
SEA	310p	503p	1927
SEA	1010p	1159p	1927
SFO	935a	1138a	2079
SFO	1225p	231p	2079
SFO	650p	849p	2079
STL	720a	752a	440
STL	1205p	1245p	440
STL	655p	735p	440
TVC	900a	1005a	207
TVC	310p	404p	207
TVC	1005p	1100p	207
TYS	700a	900a	443
TYS	700p	905p	443

Time Zone Map of the United States

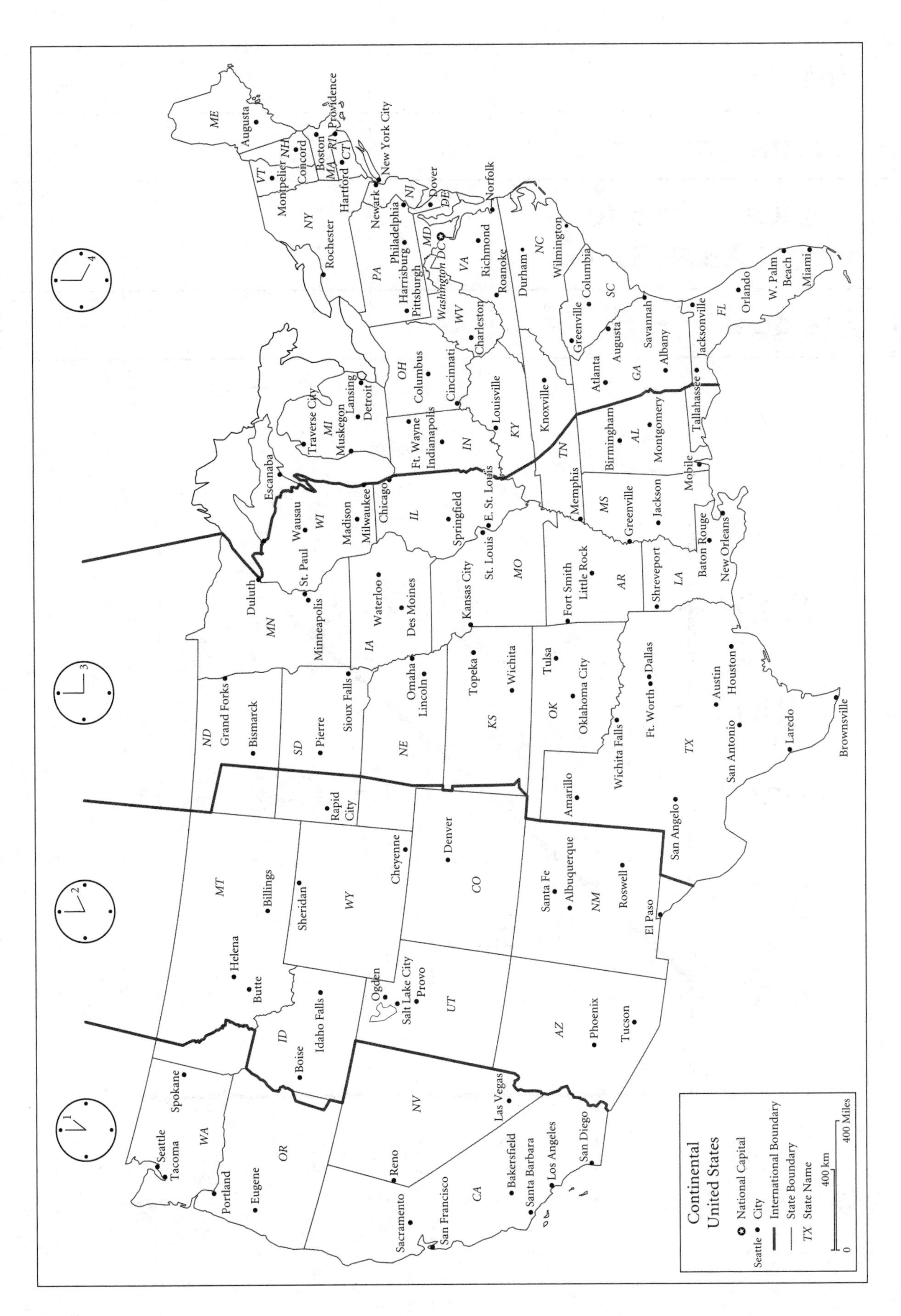

Pizza Supplies

Year (x)	1	2	3	4
Number of pizzas (y)	512	603	642	775
y-value from line $\hat{y} \approx 417.3 + 87.7x$				
Residual ($y - \hat{y}$)				

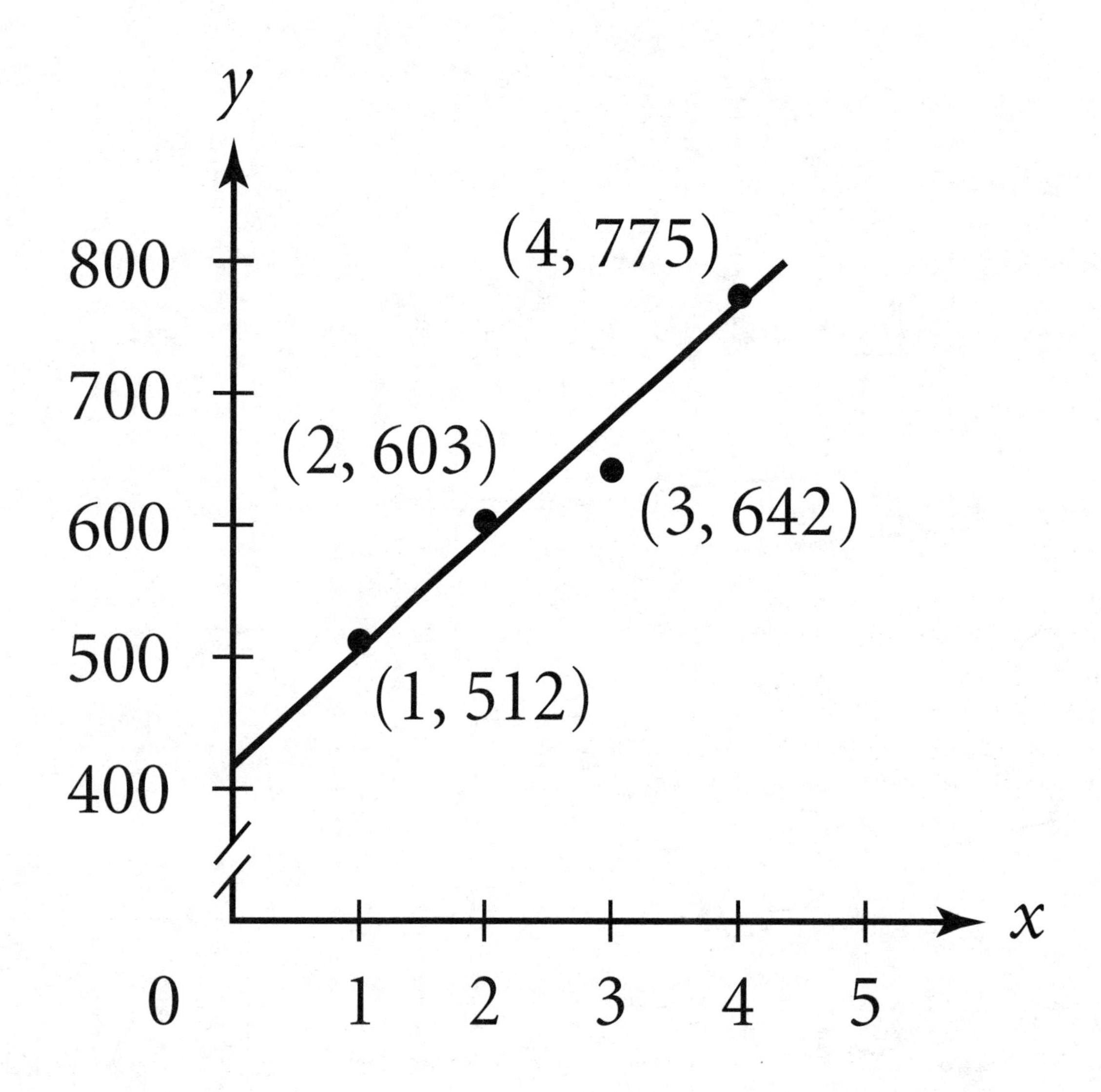

Spring Experiment Sample Data

Mass (g) x	Spring length (cm) y
50	5.0
60	5.5
70	6.0
80	6.3
90	6.8
100	7.1
110	7.5
120	7.7
130	8.1
140	8.5
150	8.8
160	9.2
170	9.5
180	9.9
190	10.3
200	10.7
210	11.3
220	11.4
230	11.9
240	12.4

Equations of Parallel and Perpendicular Lines

Parallel Slope Property In a coordinate plane, two distinct lines are parallel if and only if their slopes are equal.

Perpendicular Slope Property In a coordinate plane, two nonvertical lines are perpendicular if and only if their slopes are negative reciprocals of each other.

EXAMPLE

Consider $A(-15, -6)$, $B(6, 8)$, $C(4, -2)$, and $D(-4, 10)$.

a. Determine whether $\overleftrightarrow{AB}$ and $\overleftrightarrow{CD}$ are parallel, perpendicular, or neither.

b. Find a point E such that $\overleftrightarrow{EC} \parallel \overleftrightarrow{DB}$.

▶ ***Solution***

a. Calculate the slopes of $\overleftrightarrow{AB}$ and $\overleftrightarrow{CD}$.

$$\text{slope of } \overleftrightarrow{AB} = \frac{8 - (-6)}{6 - (-15)} = \frac{2}{3} \qquad \text{slope of } \overleftrightarrow{CD} = \frac{10 - (-2)}{-4 - 4} = -\frac{3}{2}$$

The slopes, $\frac{2}{3}$ and $-\frac{3}{2}$, are negative reciprocals of each other. Therefore $\overleftrightarrow{AB} \perp \overleftrightarrow{CD}$.

b. Find the slope of $\overleftrightarrow{DB}$.

$$\text{slope of } \overleftrightarrow{DB} = \frac{10 - 8}{-4 - 6} = -\frac{1}{5}$$

If $\overleftrightarrow{EC} \parallel \overleftrightarrow{DB}$, then the slope of $\overleftrightarrow{EC}$ equals the slope of $\overleftrightarrow{DB}$. Use $E(x, y)$ and $C(4, -2)$ in the slope formula to get $\frac{-2 - y}{4 - x} = -\frac{1}{5}$.
The slope $-\frac{1}{5}$ can be thought of as $\frac{-1}{5}$. Treat the denominators and numerators as two separate equations, and solve for x and y.

$$\begin{aligned} 4 - x &= 5 \\ -x &= 1 \\ x &= -1 \end{aligned} \qquad\qquad \begin{aligned} -2 - y &= -1 \\ -y &= 1 \\ y &= -1 \end{aligned}$$

Thus one of the many possible solutions is $E(-1, -1)$.

More Graph Stories

1.

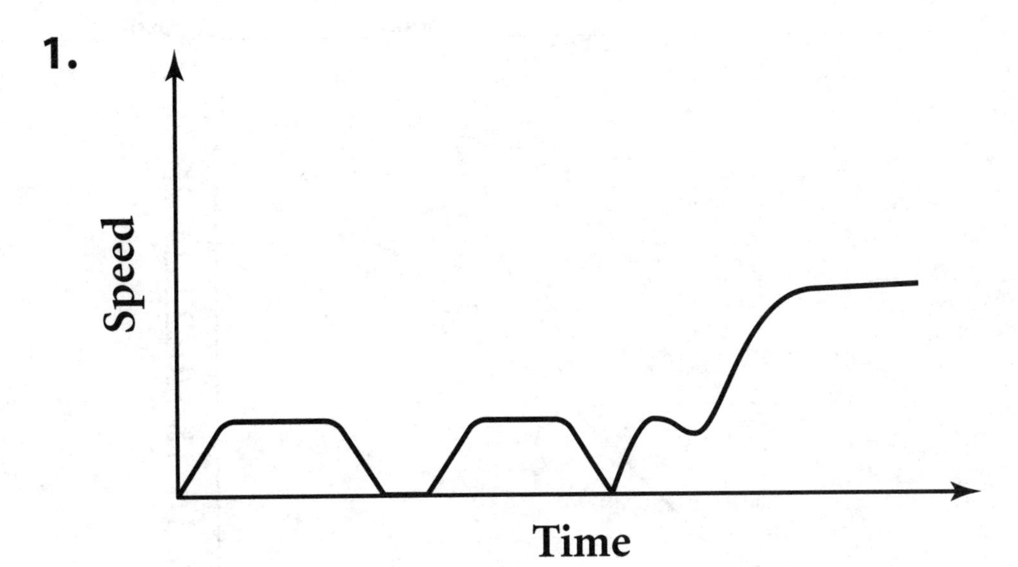

2.

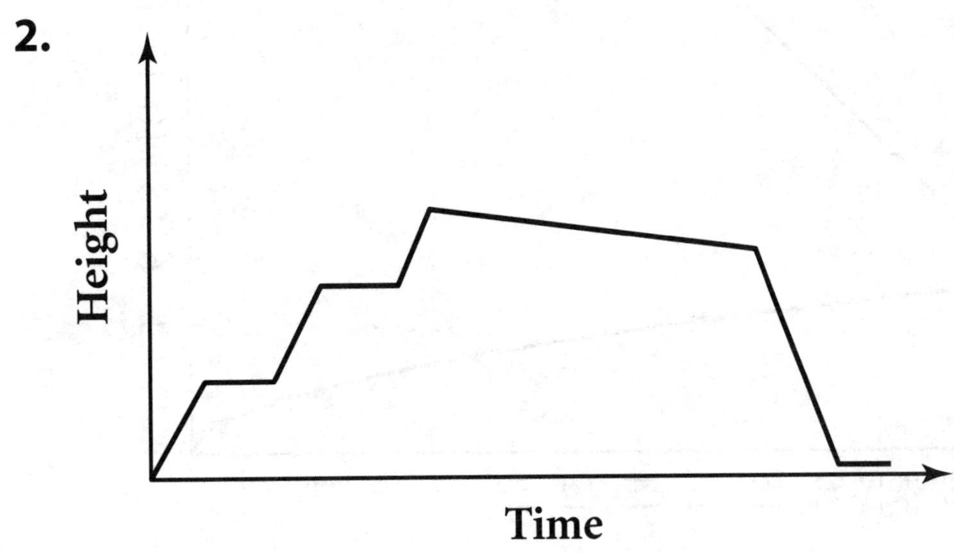

3.

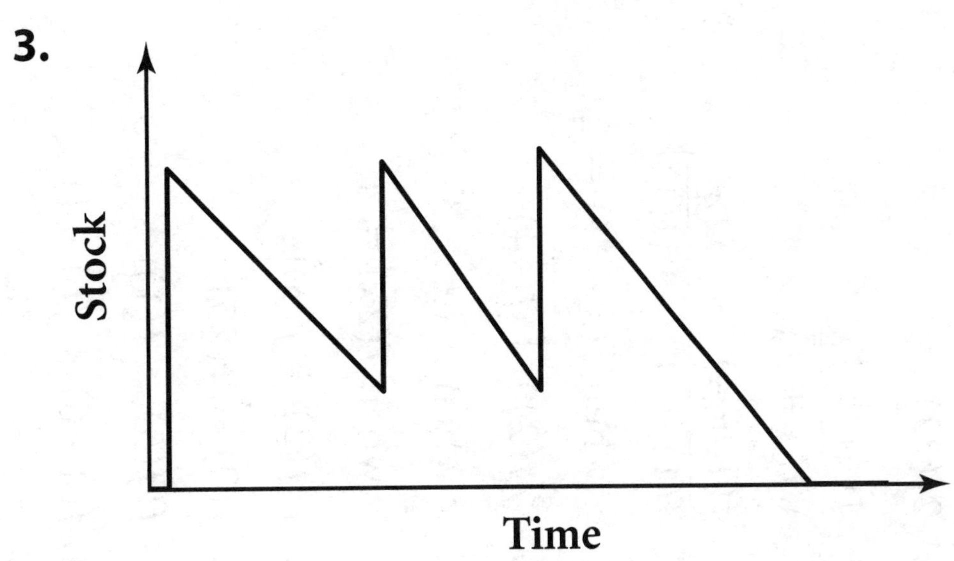

Exercise 4

Name ______________________ Period __________ Date __________

a. $f(13)$

b. $f(25) + f(26)$

c. $2f(22)$

d. $\dfrac{f(3) + 11}{\sqrt{f(3 + 1)}}$

e. $\dfrac{f(1 + 4)}{f(1) + 4} - \dfrac{1}{4}\left(\dfrac{4}{f(1)}\right)$

f. x when $f(x + 1) = 26$

g. $\sqrt[3]{f(21)} + f(14)$

h. x when $2f(x + 3) = 52$

i. x when $f(2x) = 4$

j. $f(f(2) + f(3))$

k. $f(9) - f(25)$

l. $f(f(5) - f(1))$

m. $f(4 \cdot 6) + f(4 \cdot 4)$

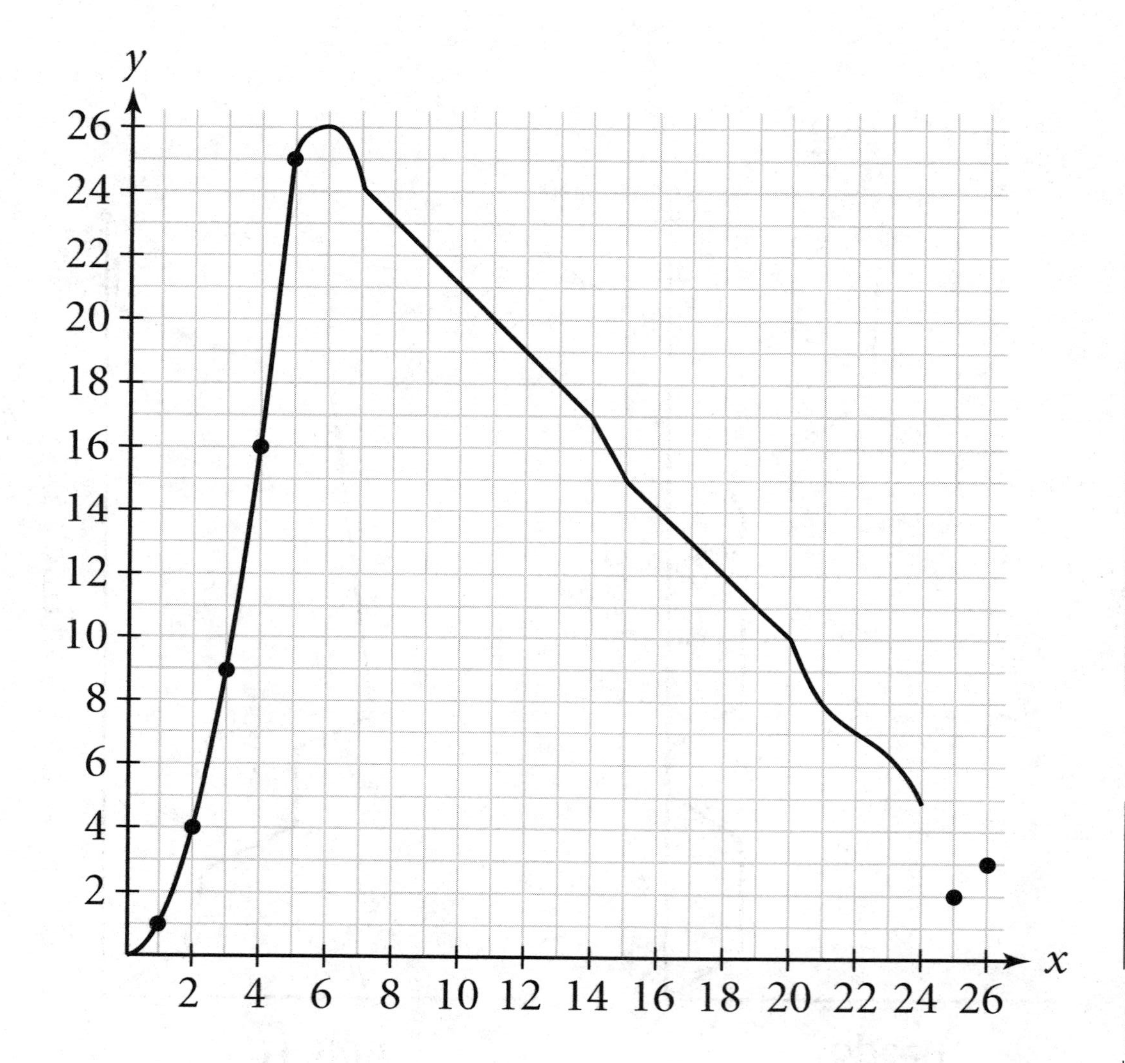

___ a ___ b ___ c ___ d ___ e ___ f ___ g ___ h ___ i ___ j ___ k ___ l ___ m

Exercise 6

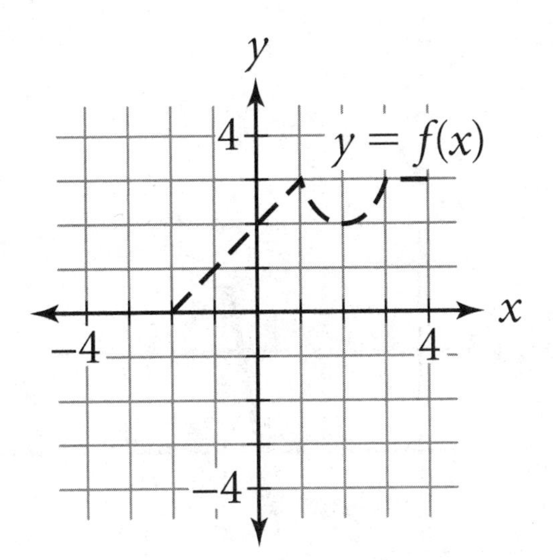

a.

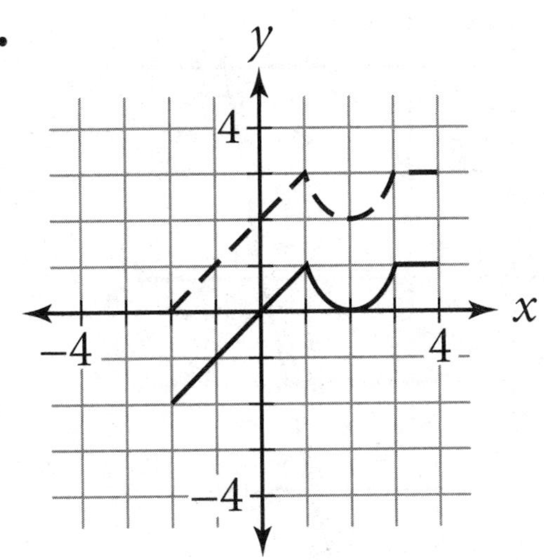

b.

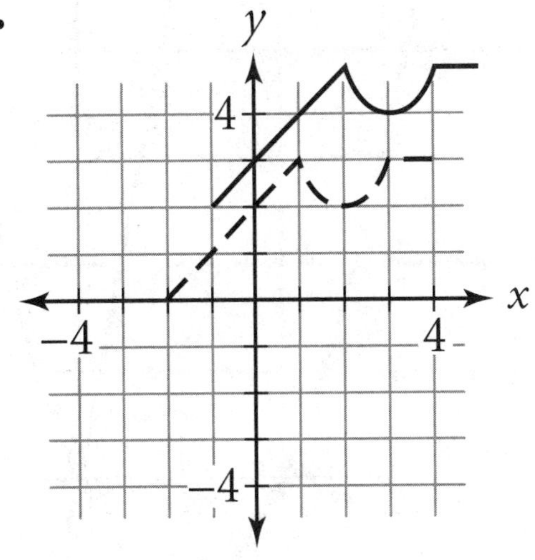

c.

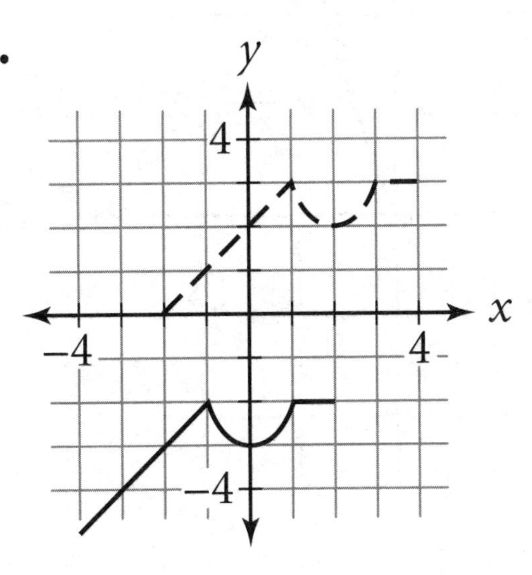

d.

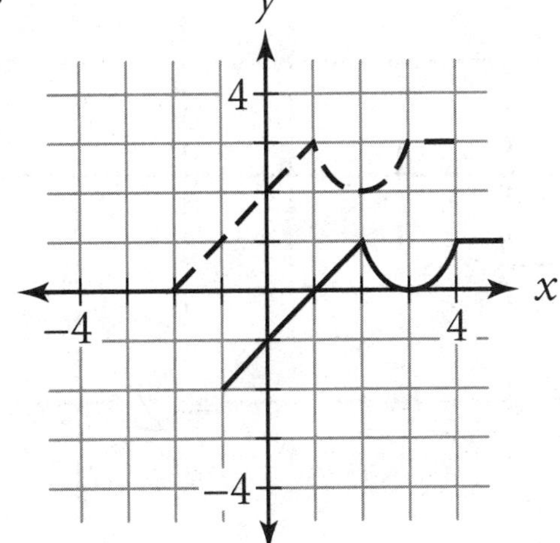

Two Parabolas

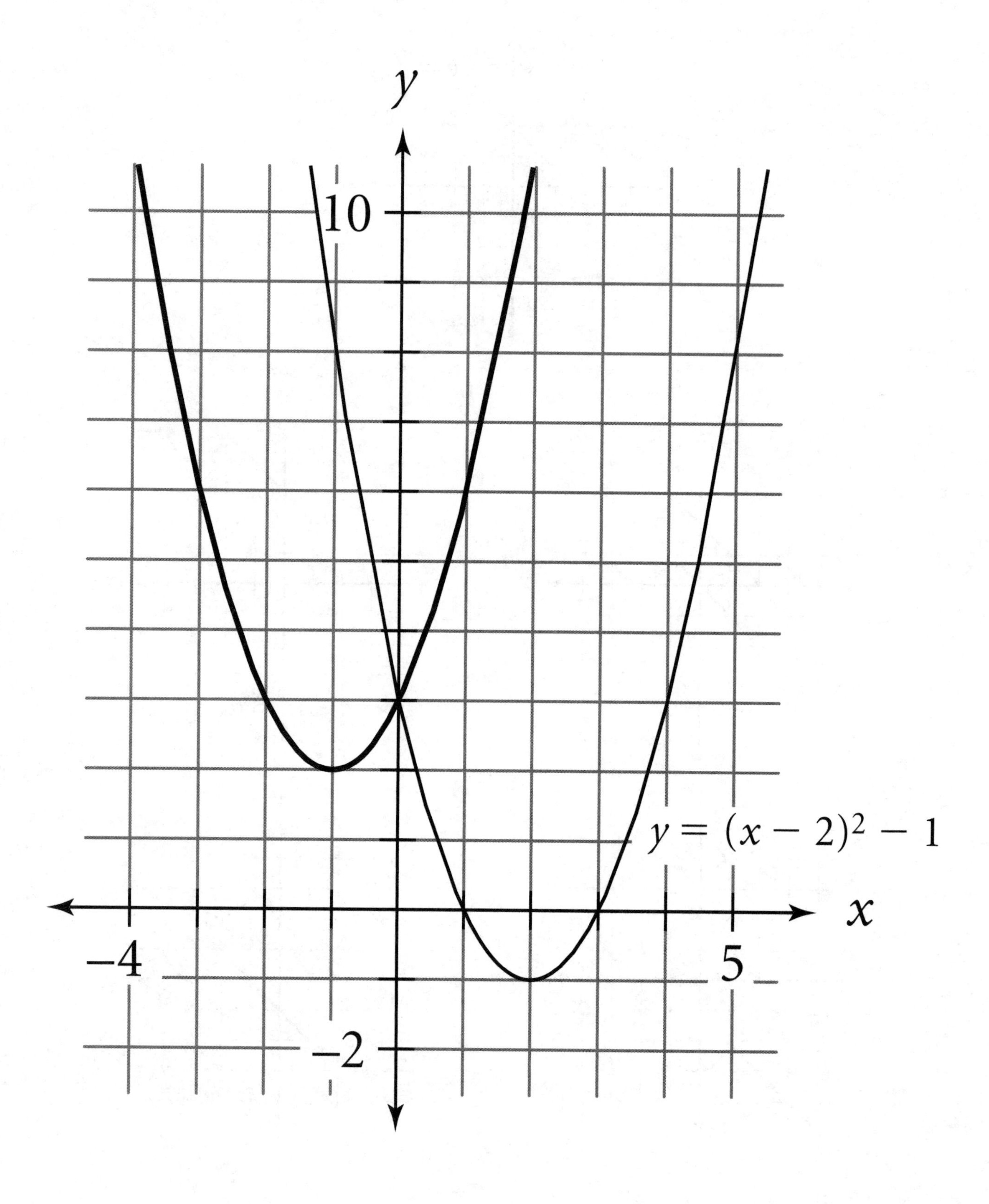

Exercise 8

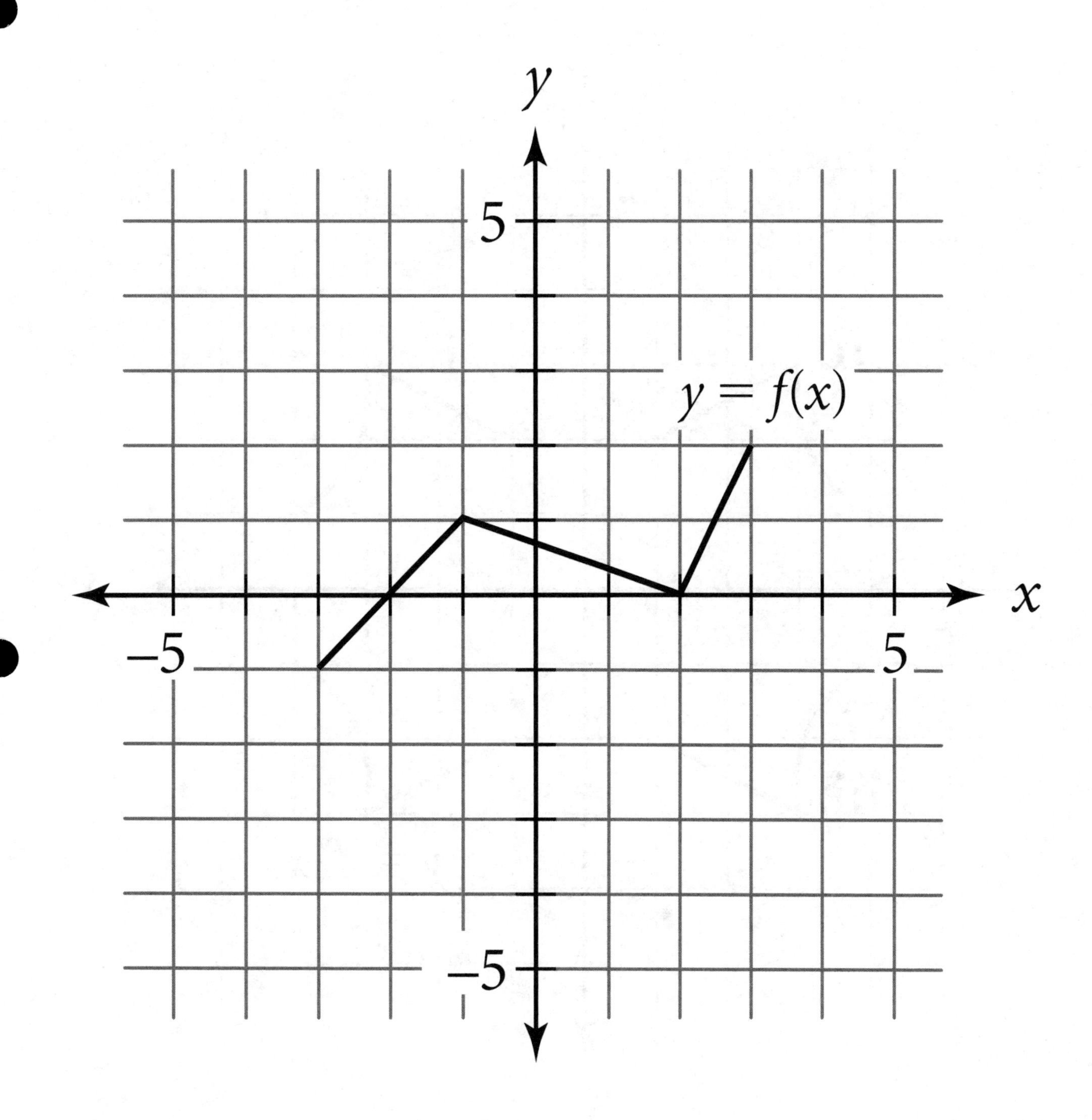

Exercise 3

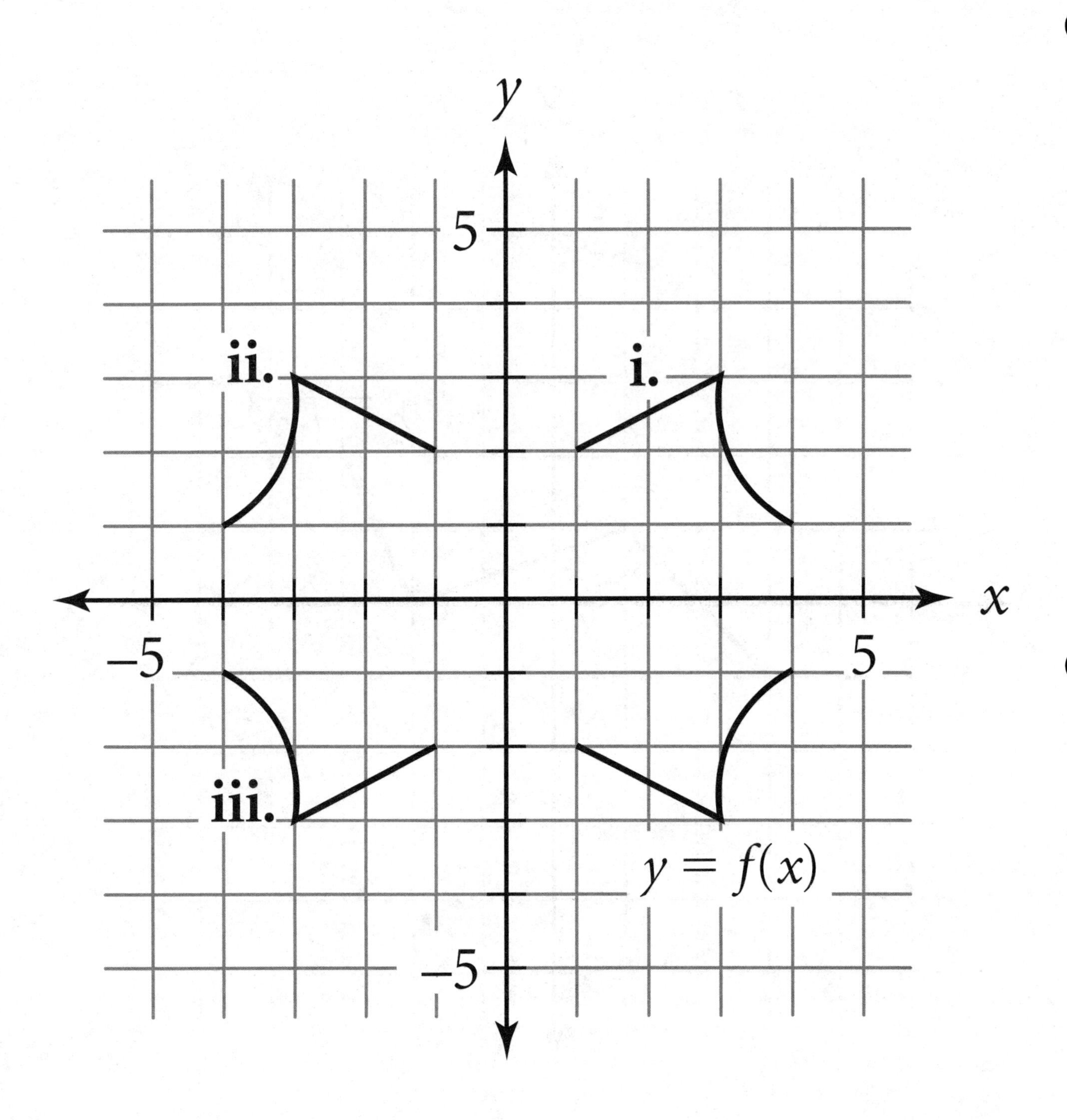

Exercise 4

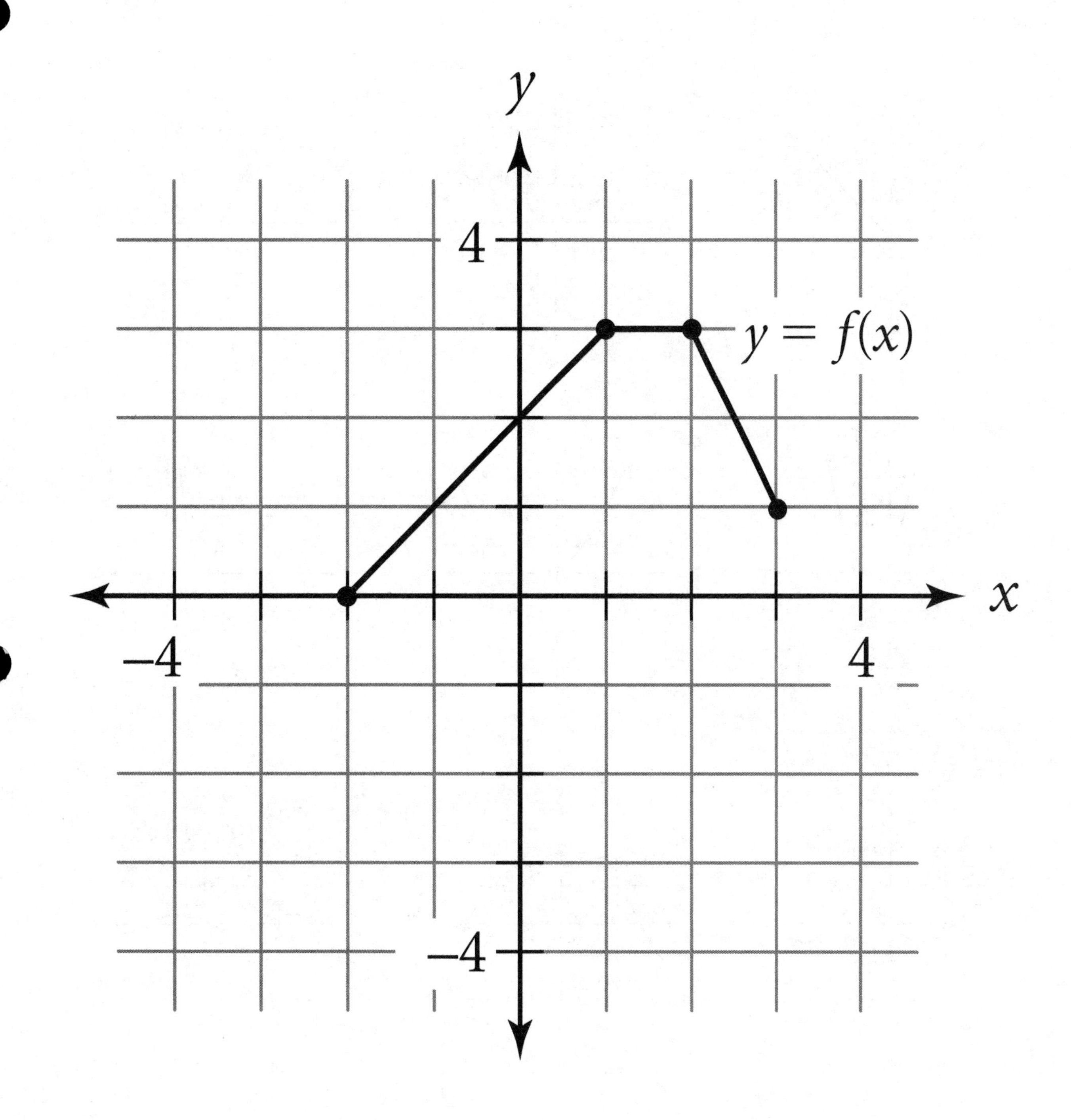

Find My Equation

$y = |x|$

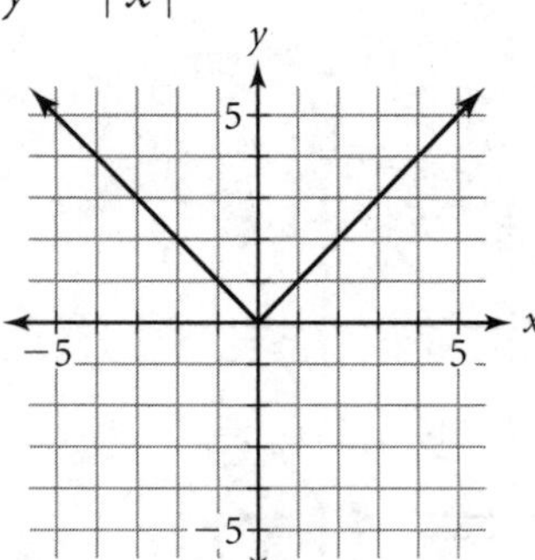

a.

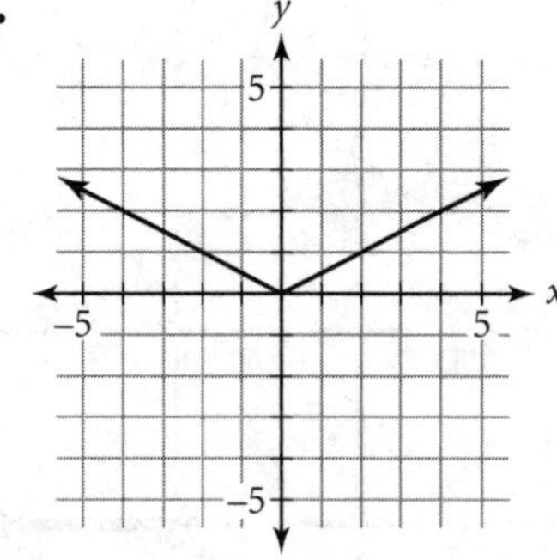

b.

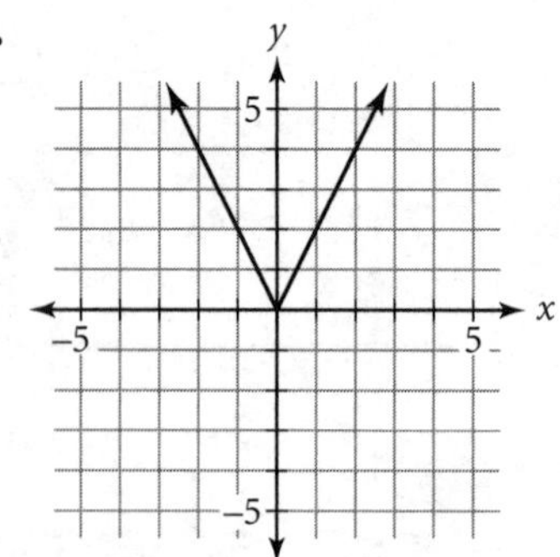

$y = (x - 1)^2$

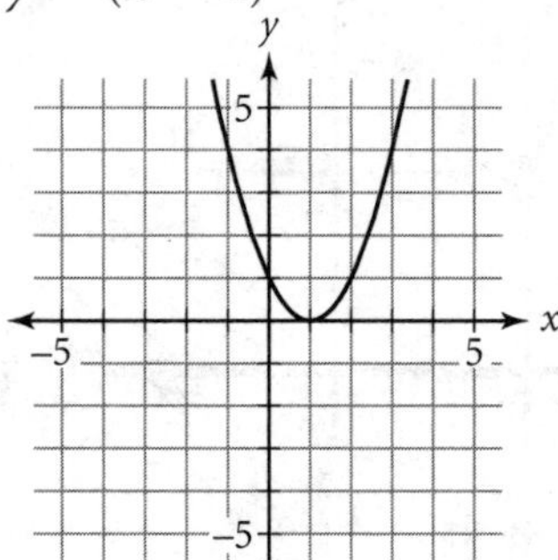

c.

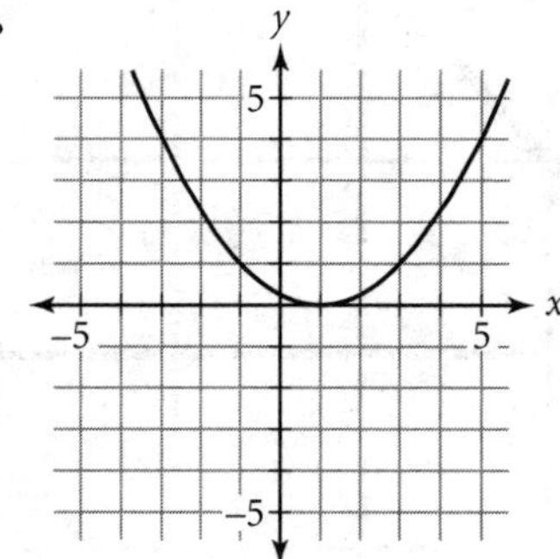

d.

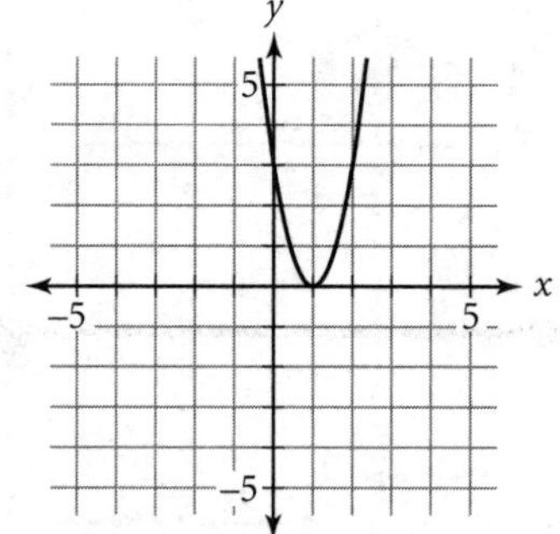

$y = f(x)$

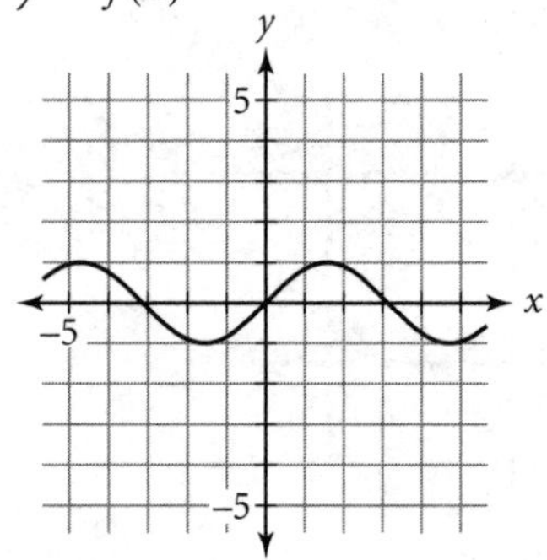

e.

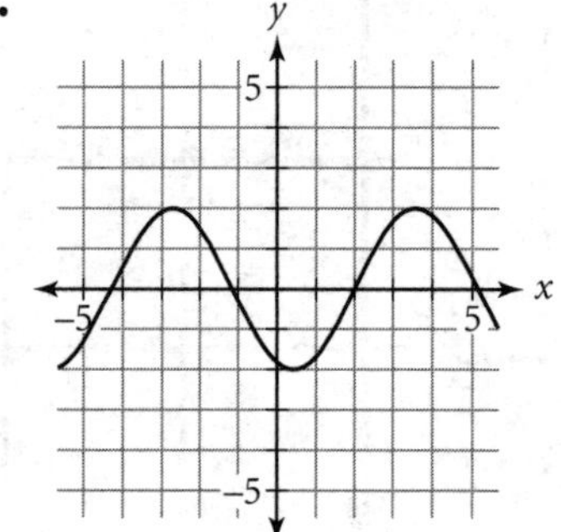

f.

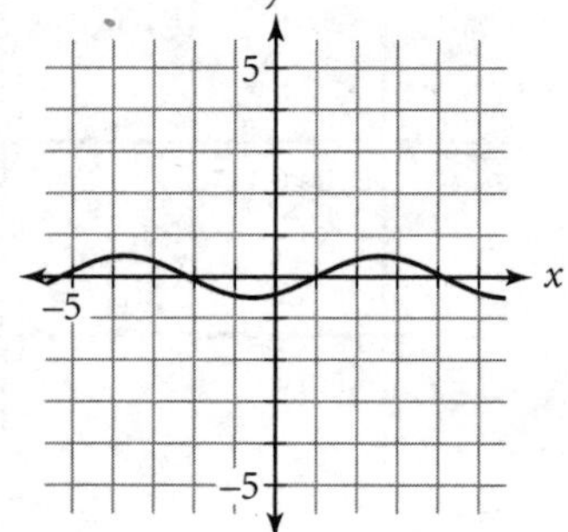

When Is a Circle Not a Circle?

Name ________________________________ Period __________ Date ______________

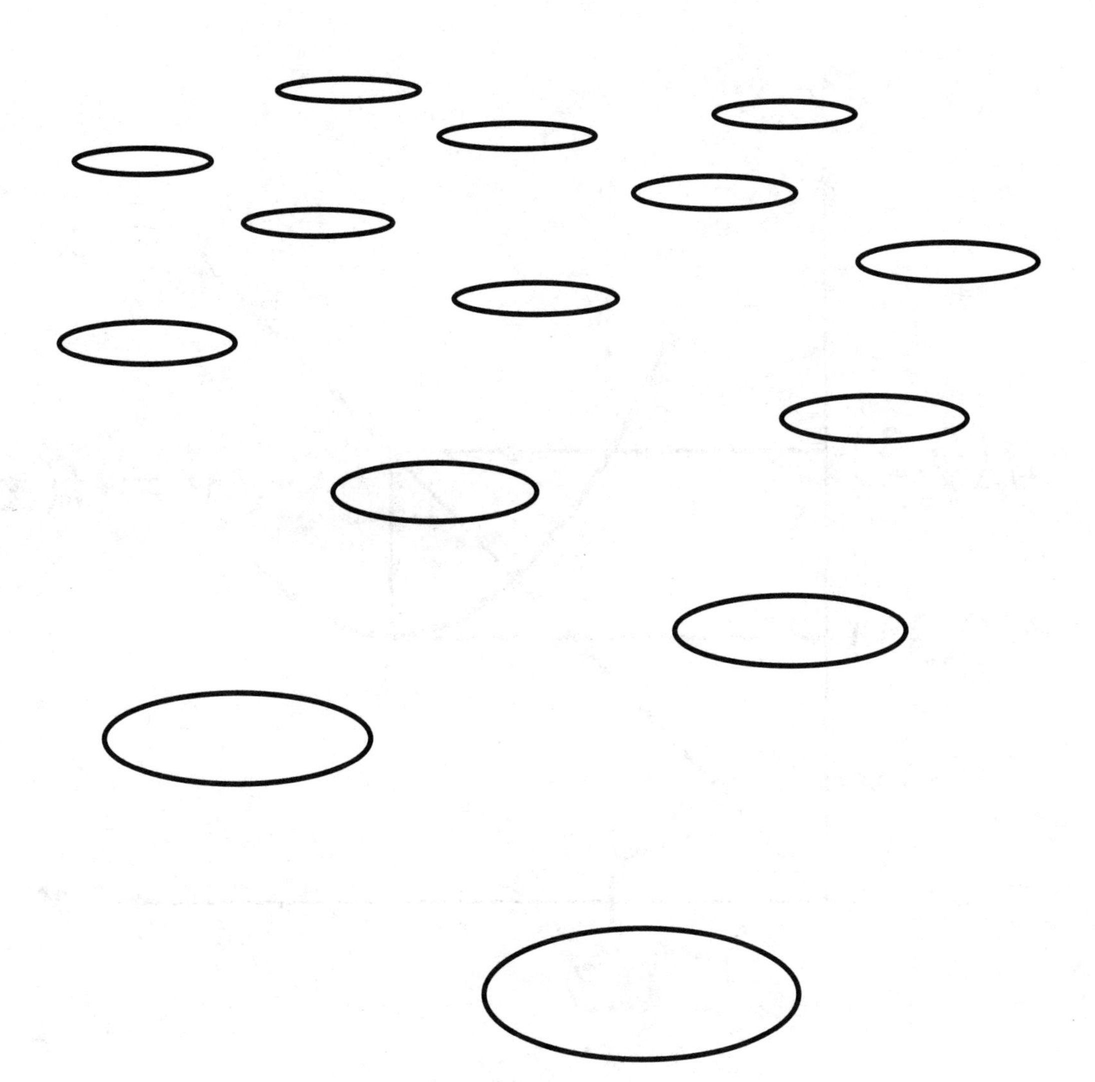

Quick Compositions?

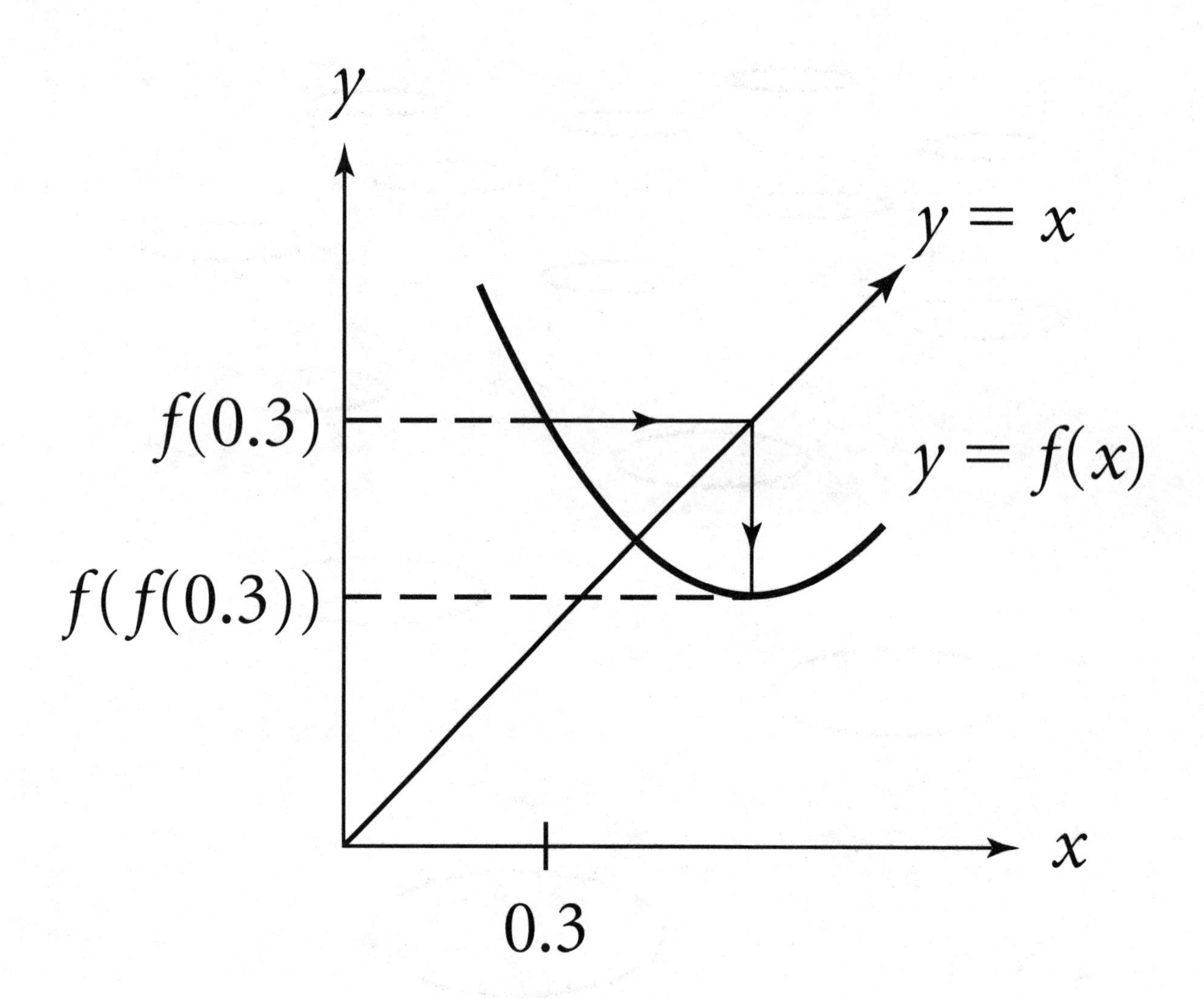

Exercises 3 and 5

Name ______________________ Period __________ Date ______________

3.

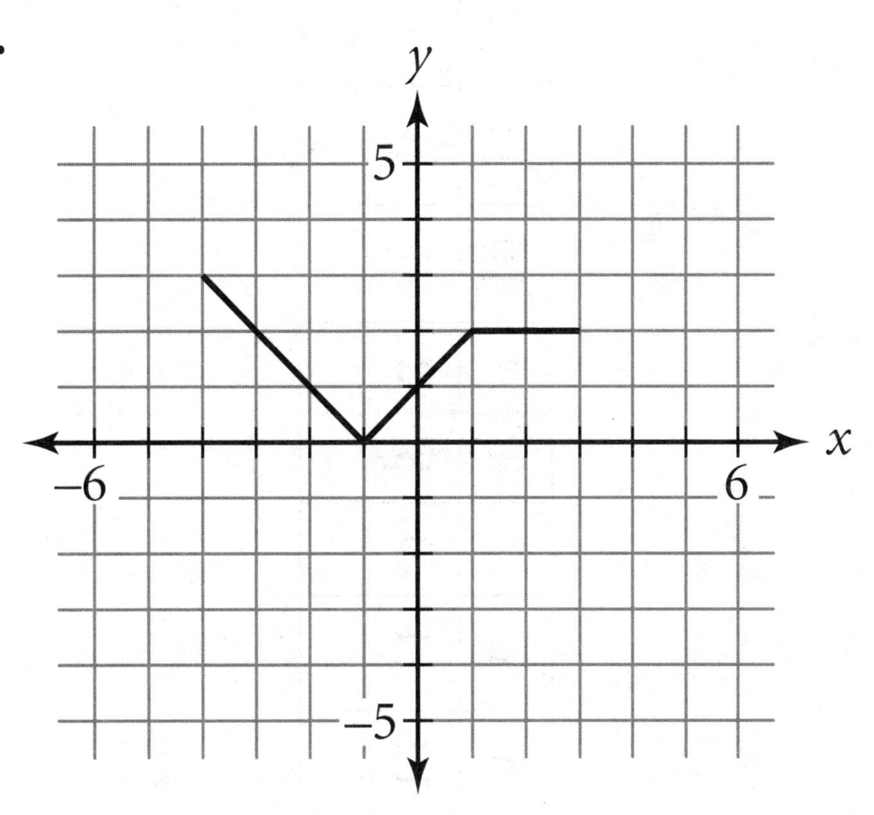

5.

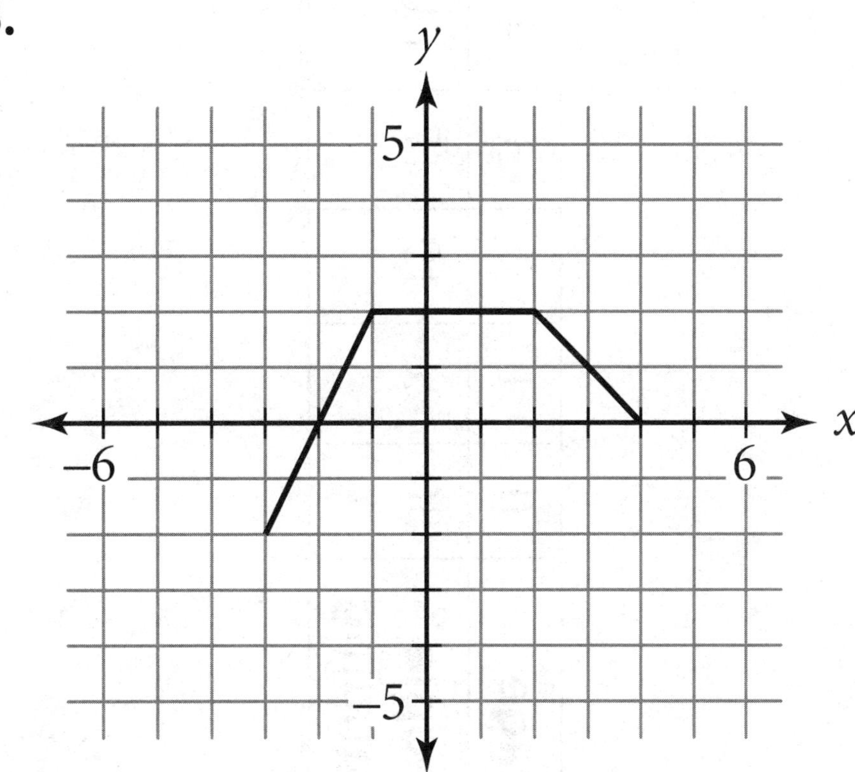

Radioactive Decay Sample Data

Stage	0	1	2	3	4	5	6	7	8	9	10	11	12
People standing	30	26	19	17	14	12	11	9	8	8	5	5	2

Pendulum Data

Swing number x	0	10	20	30	40	50	60
Closest distance (m) y	2.35	1.97	1.70	1.53	1.42	1.36	1.32

Gloria and Keith

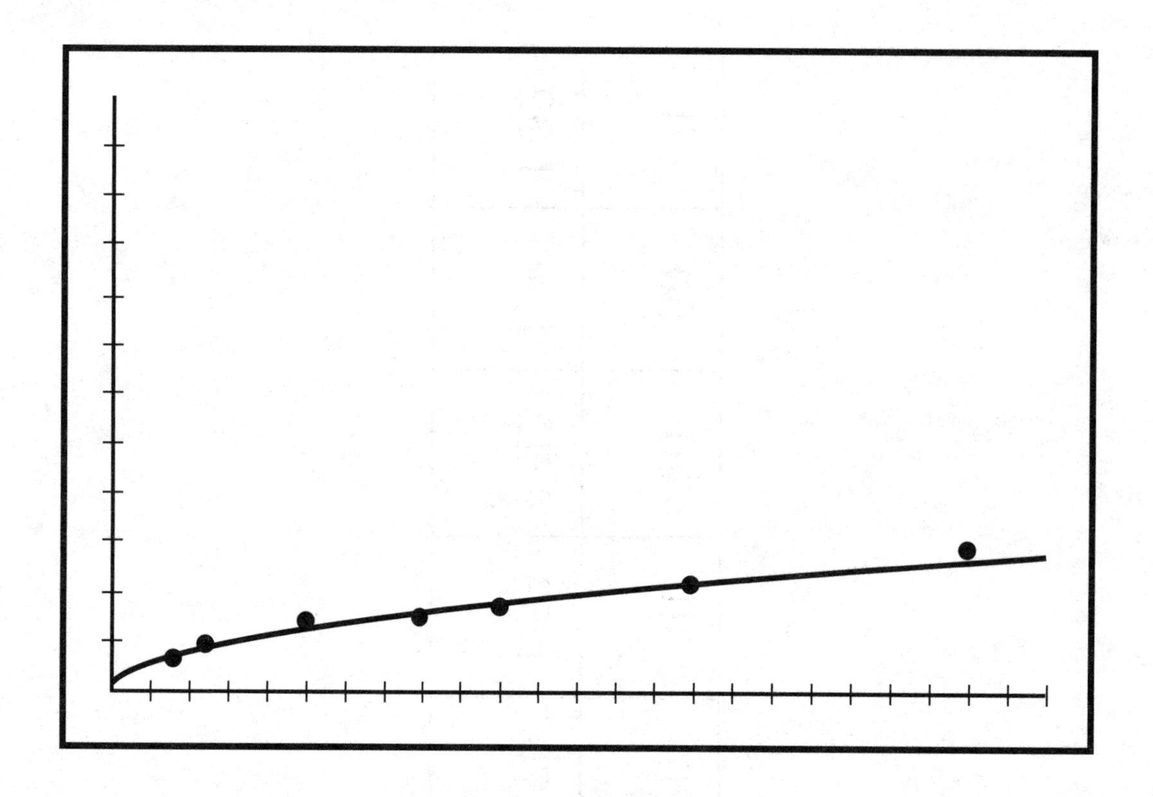

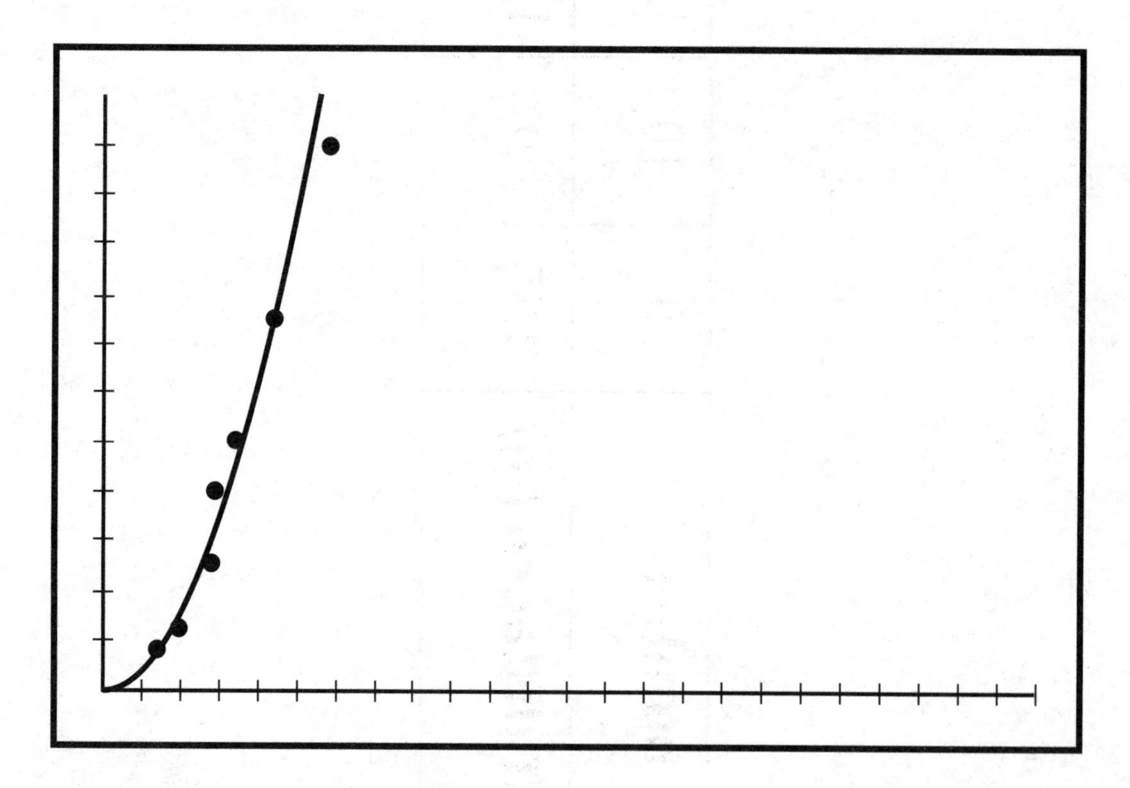

Graphing the Inverse

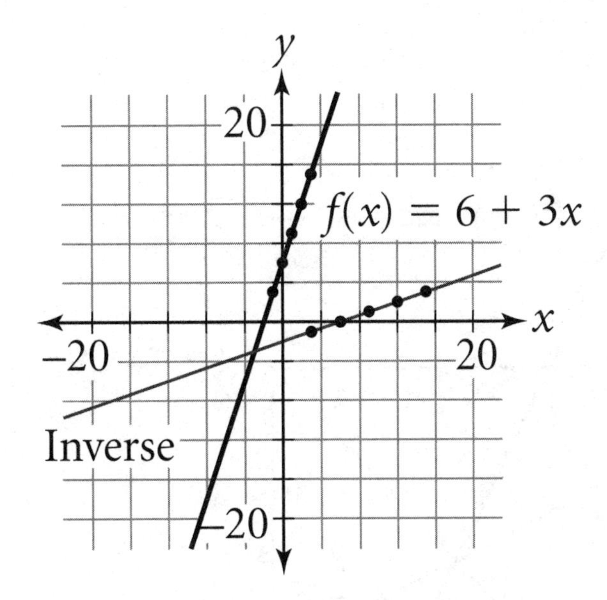

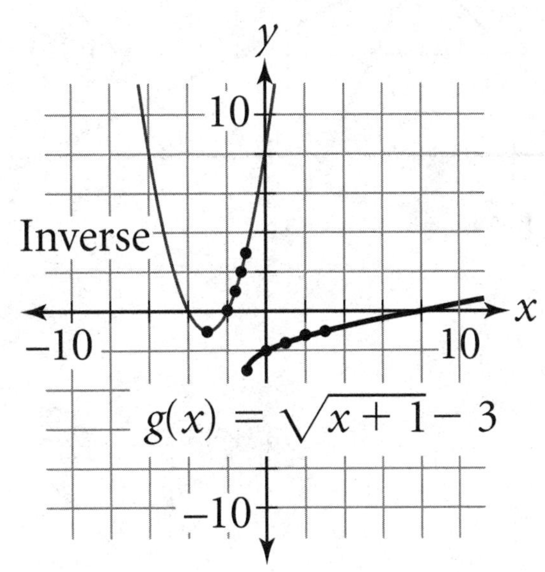

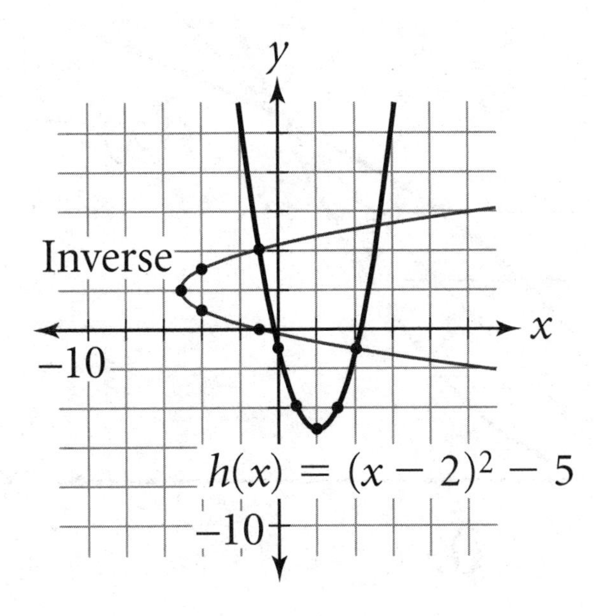

Evaluating Compositions

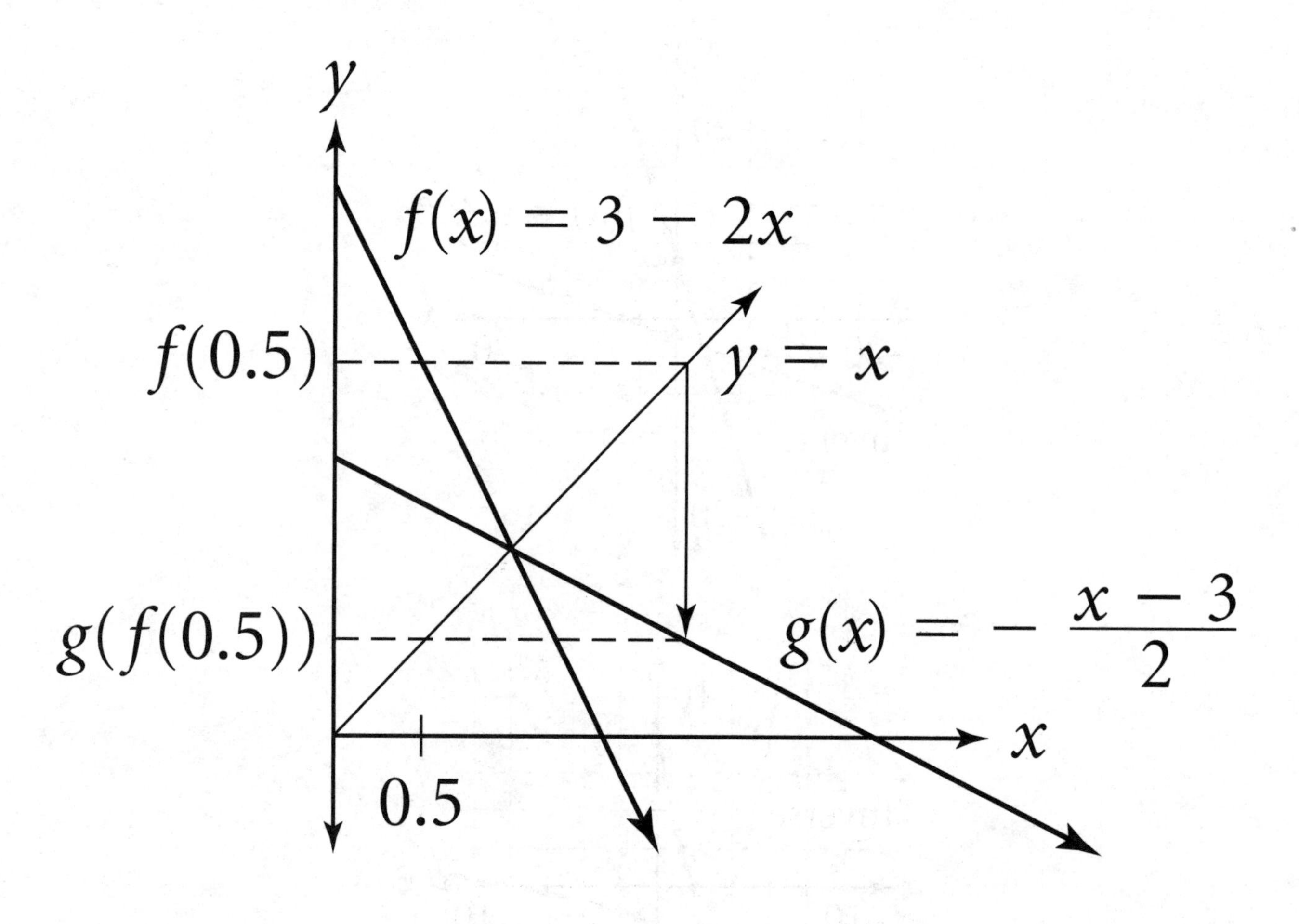

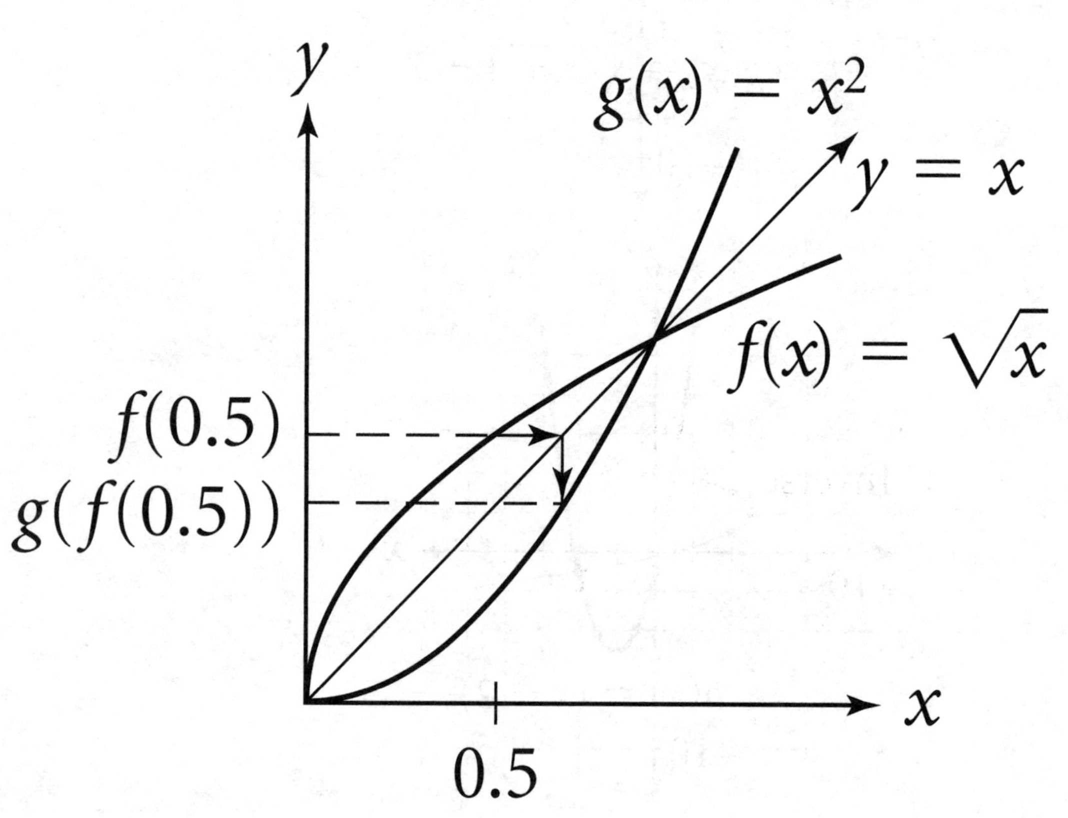

Freezer Data

Time (min)	Temperature (°C)
0	47.0
1	43.48
2	40.17
3	37.18
4	34.58
5	31.81
6	29.33
7	27.13
8	25.12
9	23.25
10	21.55

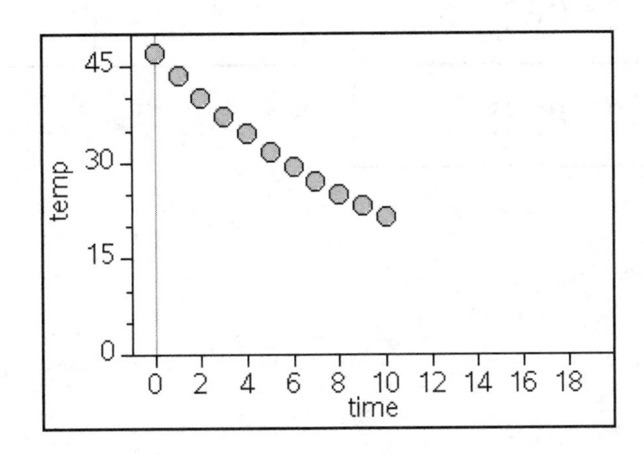

Cooling Sample Data

Time (s)	Temperature (°C)
0	29.7495
10	28.87
20	27.4995
30	26.4995
40	25.812
50	25.312
60	24.8745
70	24.437
80	24.062
90	23.8745

Time (s)	Temperature (°C)
100	23.6245
110	23.3745
120	23.187
130	22.9995
140	22.812
150	22.687
160	22.562
170	22.437
180	22.312

Cold Desserts

		Next week	
		Ice cream	Yogurt
This week	Ice cream	0.95	0.05
	Yogurt	0.10	0.90

Car Data

	City A	City B	City C
Today	24	18	11
Tomorrow	Add double today and then add 31	Add double today and then add 23	Add double today and then add 15

	Next day to A	Next day to B	Next day to C
One day from A	0.73	0.23	0.04
One day from B	0.14	0.81	0.05
One day from C	0.06	0.05	0.89

Find Your Place

Two-Vertex Model

City A	City B
$x \leq 0.3 \to$ **B** $x > 0.3$ **stay**	$x \leq 0.2 \to$ **A** $x > 0.2$ **stay**

Three-Vertex Model

City A	City B	City C
$x \leq 0.2 \to$ **B** $0.2 < x \leq 0.7 \to$ **C** $x > 0.7$ **stay**	$x \leq 0.5 \to$ **A** $x > 0.5$ **stay**	$x \leq 0.1 \to$ **B** $0.1 < x \leq 0.3 \to$ **A** $x > 0.3$ **stay**

Four-Vertex Model

City A	City B
$x \leq 0.2 \to$ **B** $0.2 < x \leq 0.6 \to$ **D** $x > 0.6 \to$ **C**	$x \leq 0.3 \to$ **A** $0.3 < x \leq 0.8 \to$ **C** $x > 0.8$ **stay**
City C	**City D**
$x \leq 0.2 \to$ **B** $0.2 < x \leq 0.5 \to$ **A** $x > 0.5$ **stay**	$x \leq 0.1 \to$ **C** $0.1 < x \leq 0.4 \to$ **B** $x > 0.4$ **stay**

At the Movies

	Duane	Marsha	Parker
Candy bars	2	1	0
Small drinks	1	2	2
Bags of peanuts	2	1	3
Total	$11.85	$9.00	$12.35

Pottery

	Unglazed birdbaths	Glazed birdbaths	Time available
Wheel	0.5 h	1 h	8 h
Kiln	3 h	18 h	60 h
Number made	6 + ___	___	

Graphing Inequalities

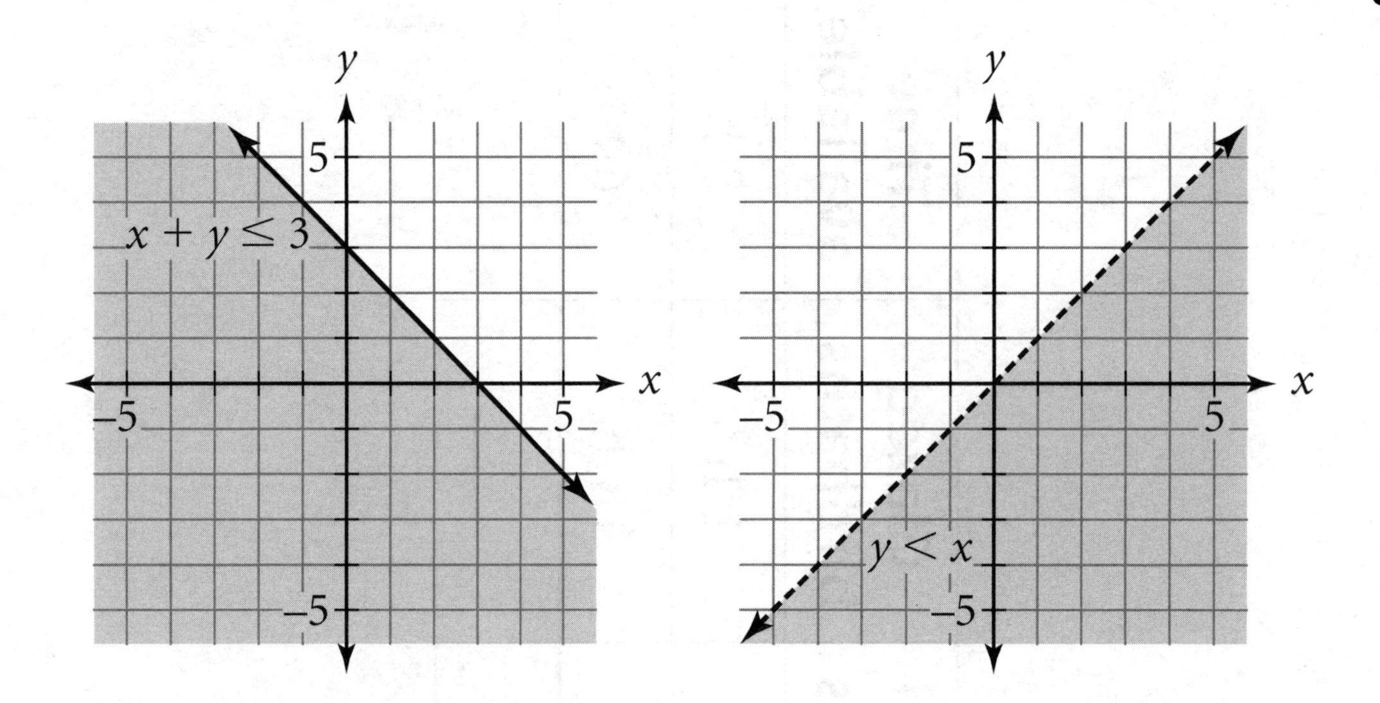

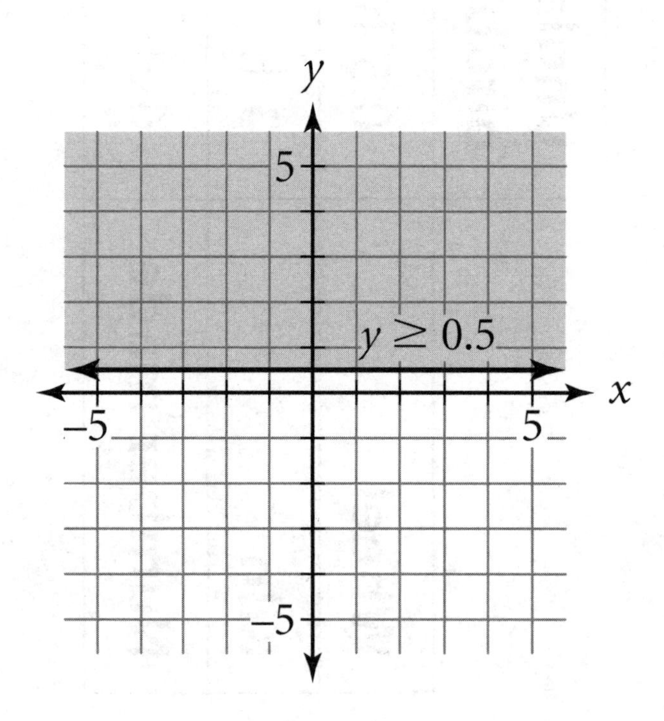

Maximizing Profit

	Amount per unglazed birdbath	Amount per glazed birdbath	Constraining value
Wheel hours			
Kiln hours			
Profit			Maximize

Snack Time

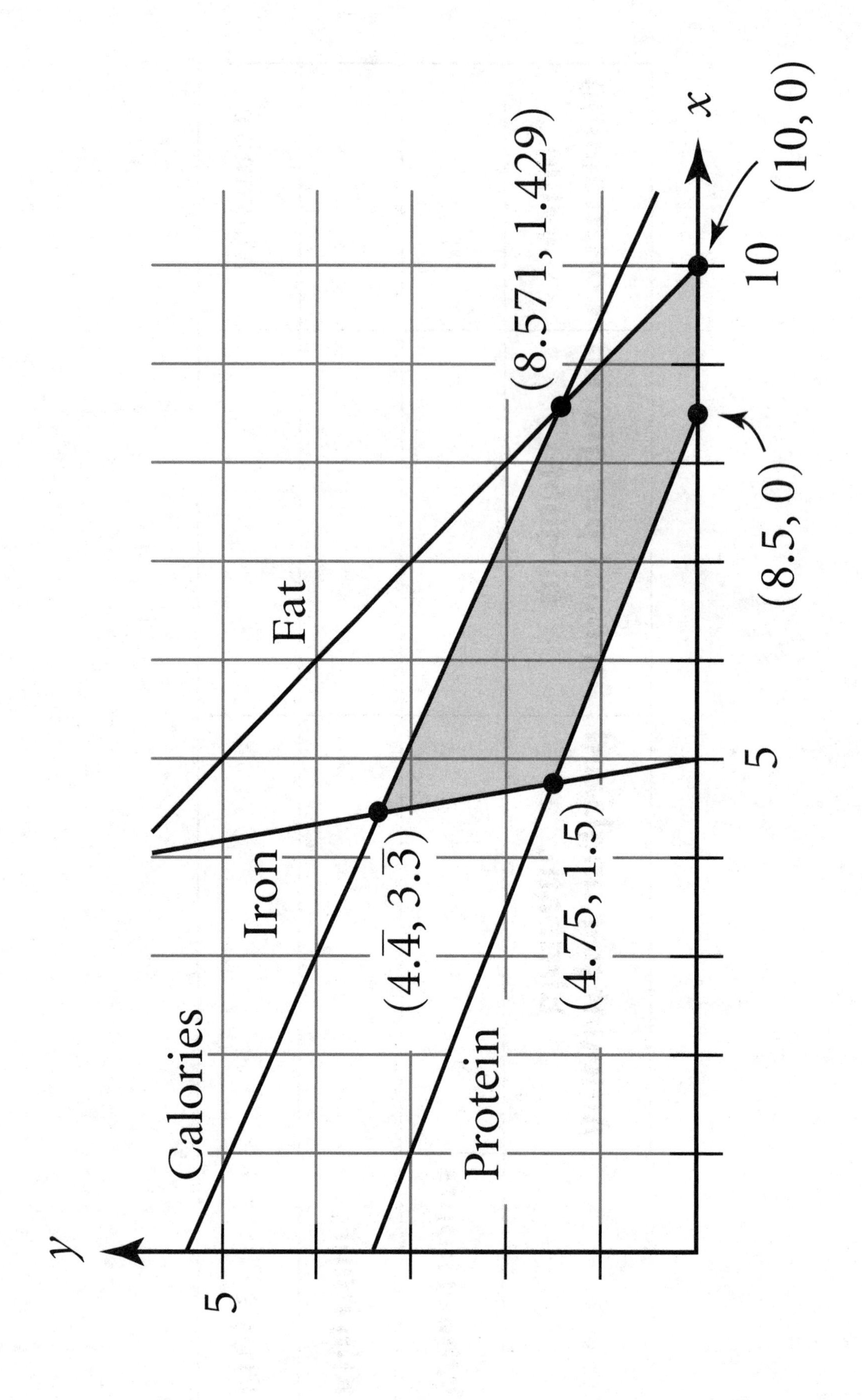

y
x
5
5
10
Calories
Iron
Fat
Protein
$(4.\overline{4}, 3.\overline{3})$
(4.75, 1.5)
(8.571, 1.429)
(8.5, 0)
(10, 0)

Free Fall Sample Data

Time (s) x	Height (m) y
0.00	2.000
0.05	1.988
0.10	1.951
0.15	1.890
0.20	1.804
0.25	1.694
0.30	1.559
0.35	1.400
0.40	1.216
0.45	1.008

Finite Differences

1st-degree polynomial

$y = 3x + 4$

x	y	D_1
2	10	
3	13	
4	16	
5	19	
6	22	
7	25	

2nd-degree polynomial

$y = 2x^2 - 5x - 7$

x	y	D_1	D_2
3.7	1.88		
3.8	2.88		
3.9	3.92		
4.0	5.00		
4.1	6.12		
4.2	7.28		

3rd-degree polynomial

$y = 0.1x^3 - x^2 + 3x - 5$

x	y	D_1	D_2	D_3
−5	−57.5			
0	−5			
5	−2.5			
10	25			
15	152.5			
20	455			

Rolling Along Sample Data

Time (s)	Distance from motion sensor (m)
0.2	0.143
0.4	0.145
0.6	0.143
0.8	0.316
1.0	1.046
1.2	1.720
1.4	2.321
1.6	2.857
1.8	3.325
2.0	3.731
2.2	4.070
2.4	4.341
2.6	4.548
2.8	4.688
3.0	4.756

Time (s)	Distance from motion sensor (m)
3.2	4.757
3.4	4.693
3.6	4.562
3.8	4.371
4.0	4.119
4.2	3.809
4.4	3.438
4.6	3.010
4.8	2.528
5.0	1.993
5.2	1.397
5.4	0.761
5.6	0.101
5.8	0.074
6.0	0.081

Quadratic List

$$f(x) = (x - 2)^2 - 5$$

$$g(x) = (x - 2)^2 - 3$$

$$h(x) = (x - 2)^2 - 1$$

$$k(x) = (x - 2)^2 + 1$$

$$p(x) = (x - 2)^2 + 3$$

$$q(x) = (x - 2)^2 + 5$$

Number System Venn Diagram

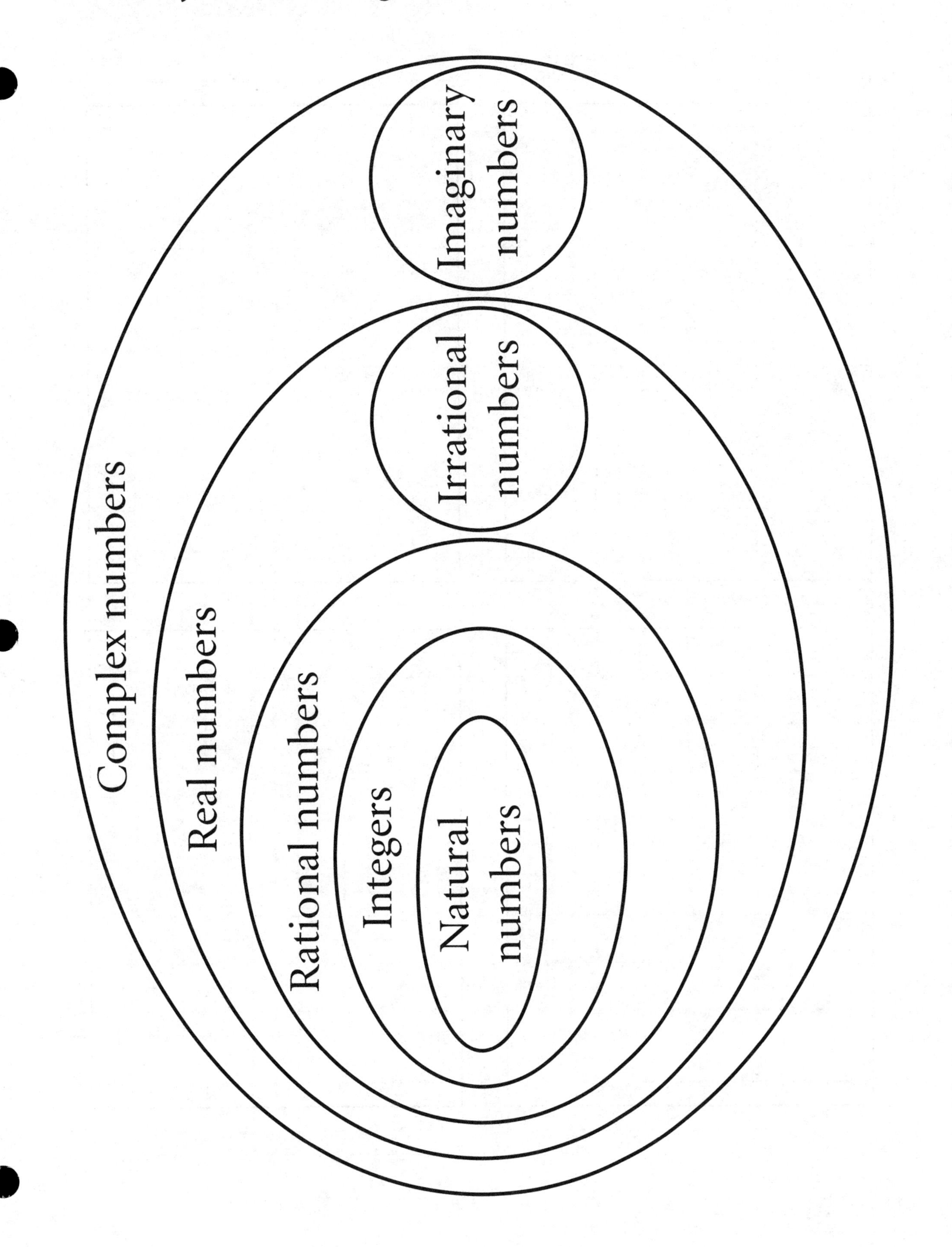

16-by-20 Grid Paper

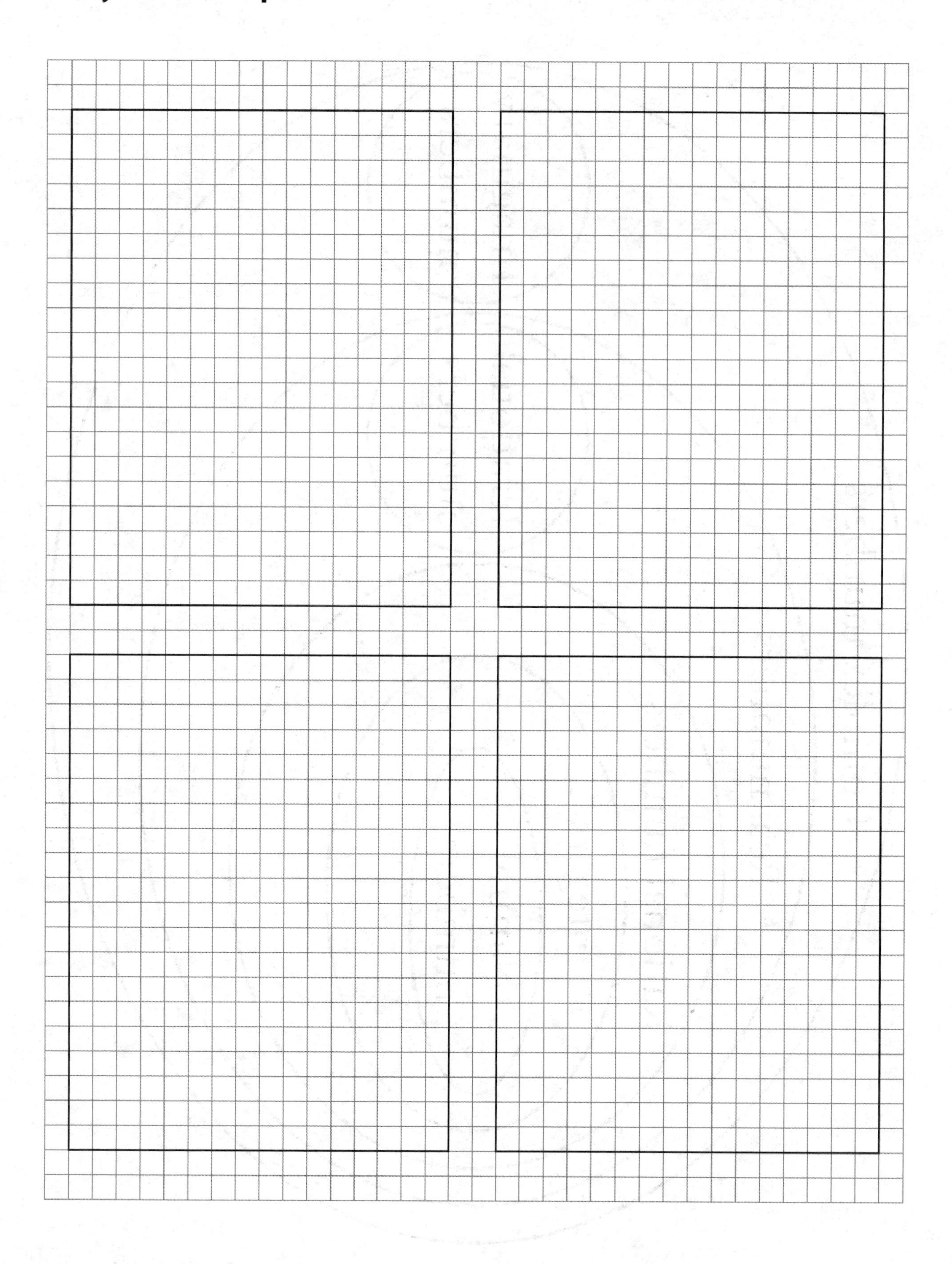

Equations and Parabolas

$y = x^2 - x + 1$

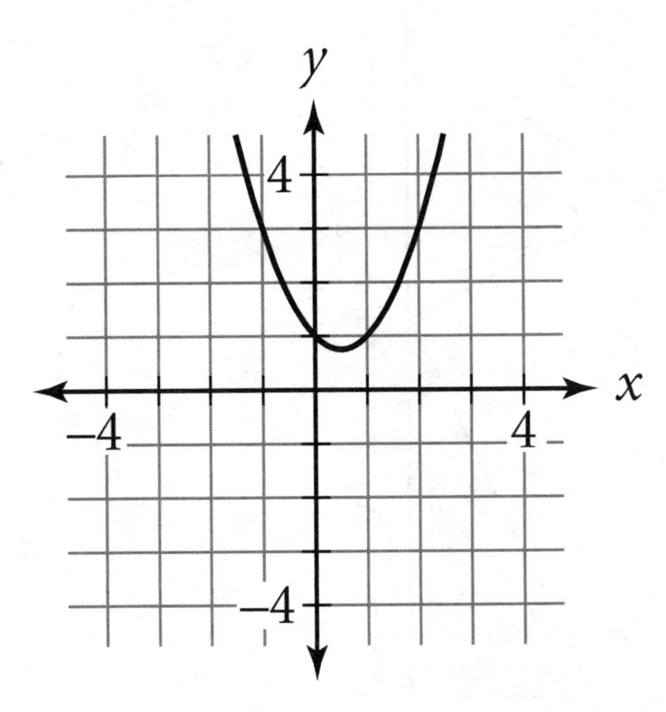

$y = x^2 - x - 1$

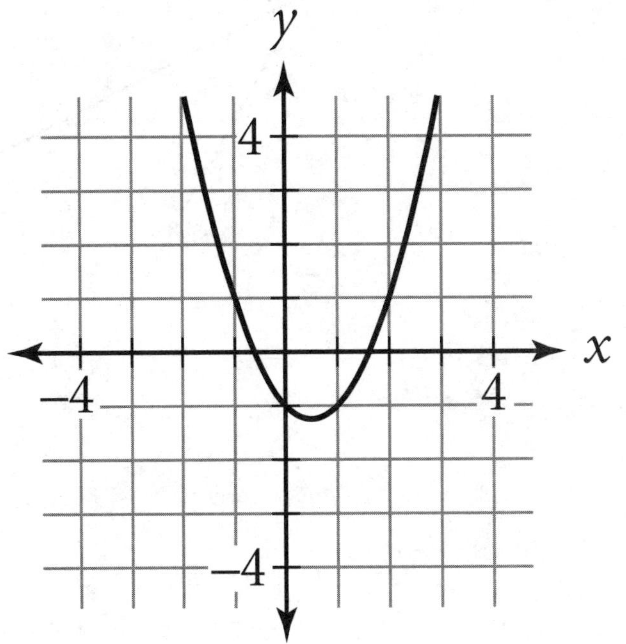

$y = x^2 - x - 2$

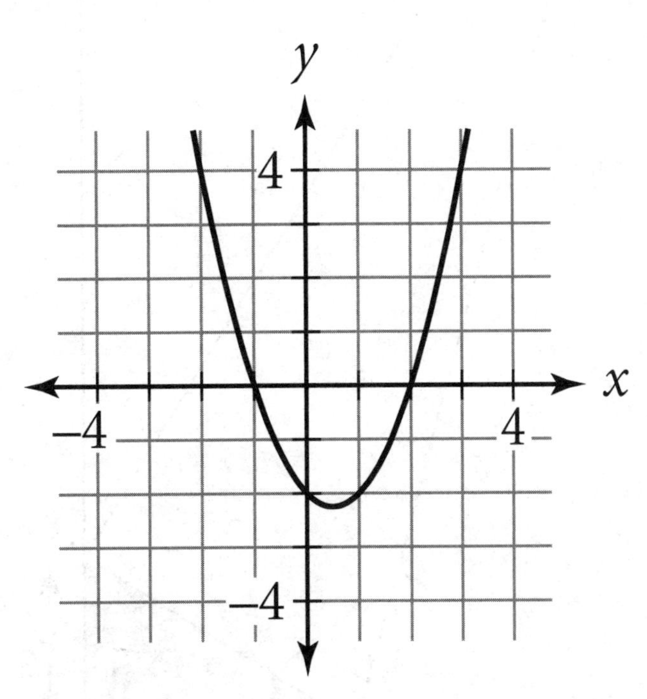

Paper Folding

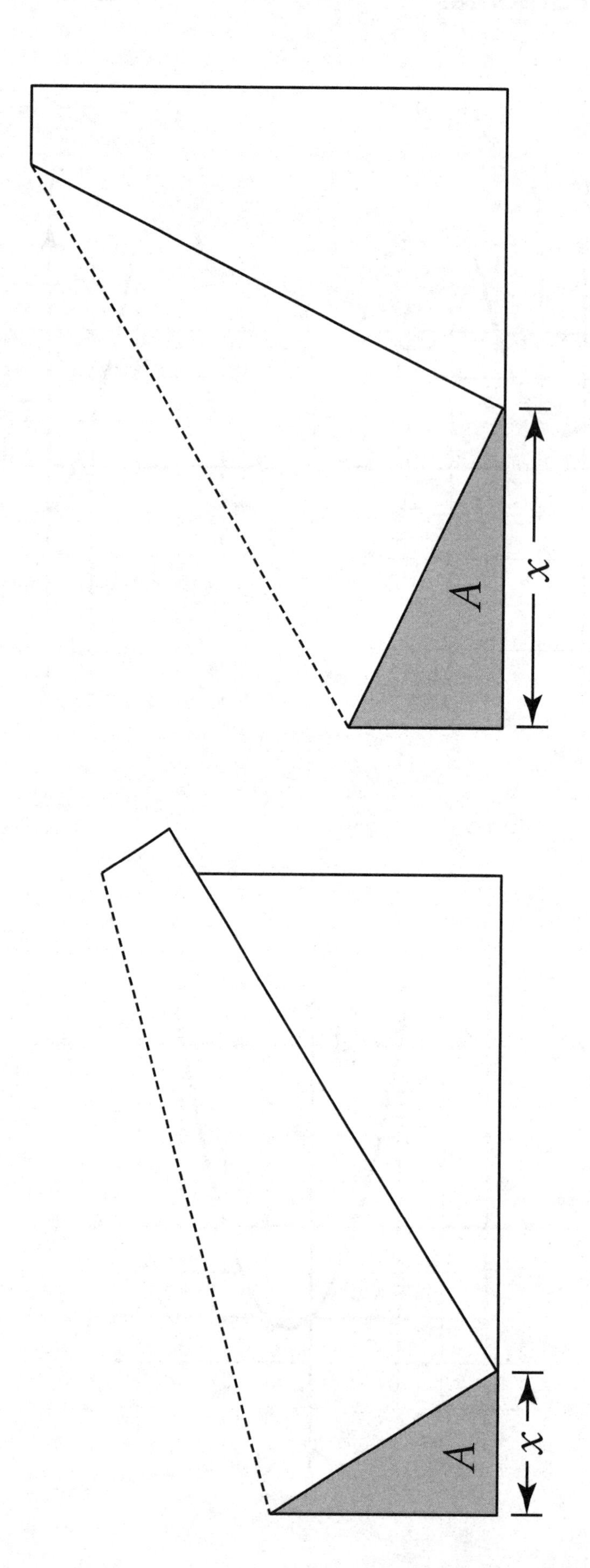

Polynomial Long Division

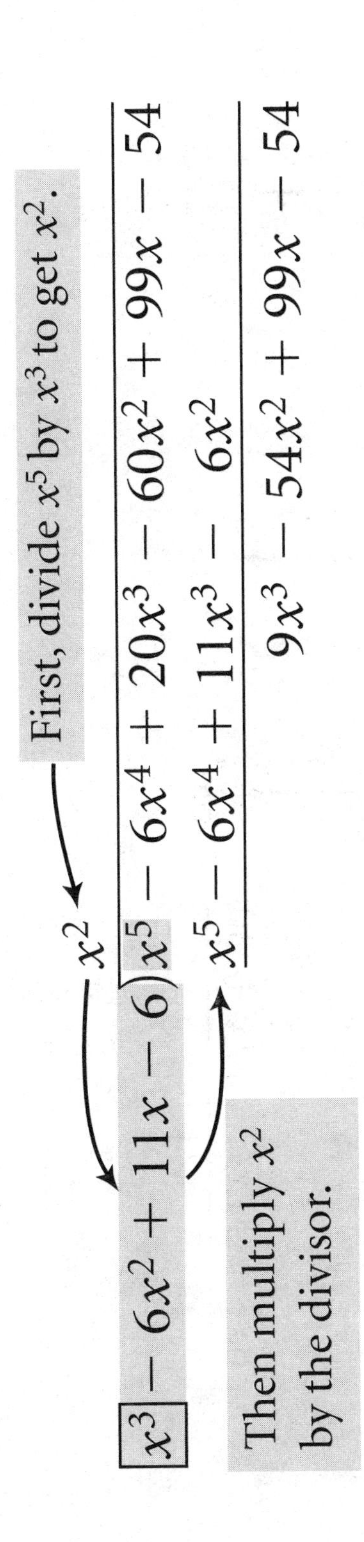

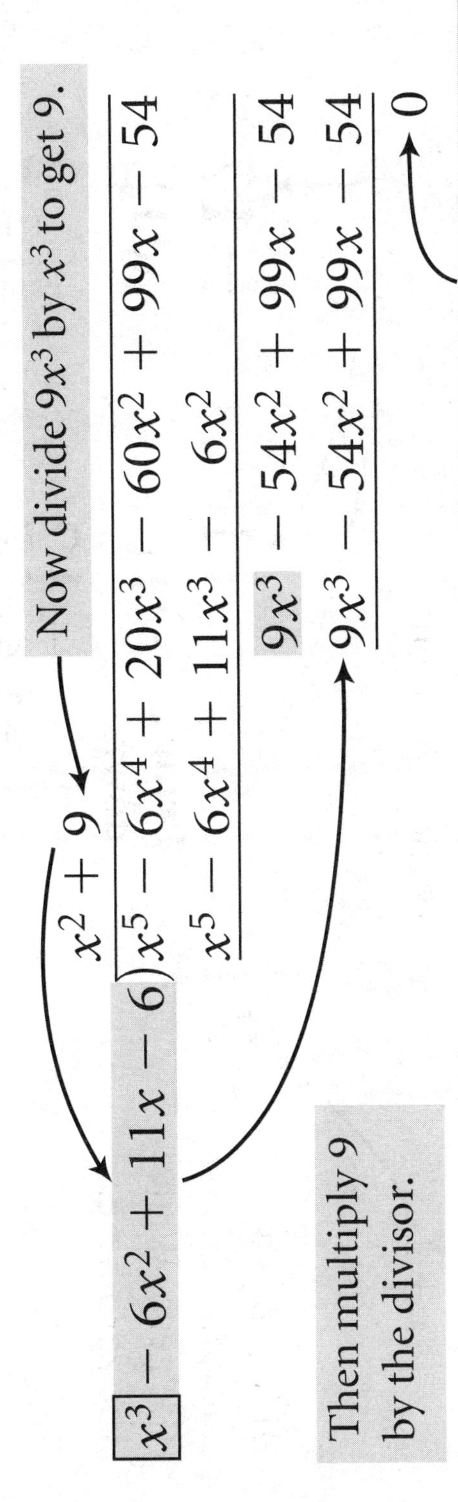

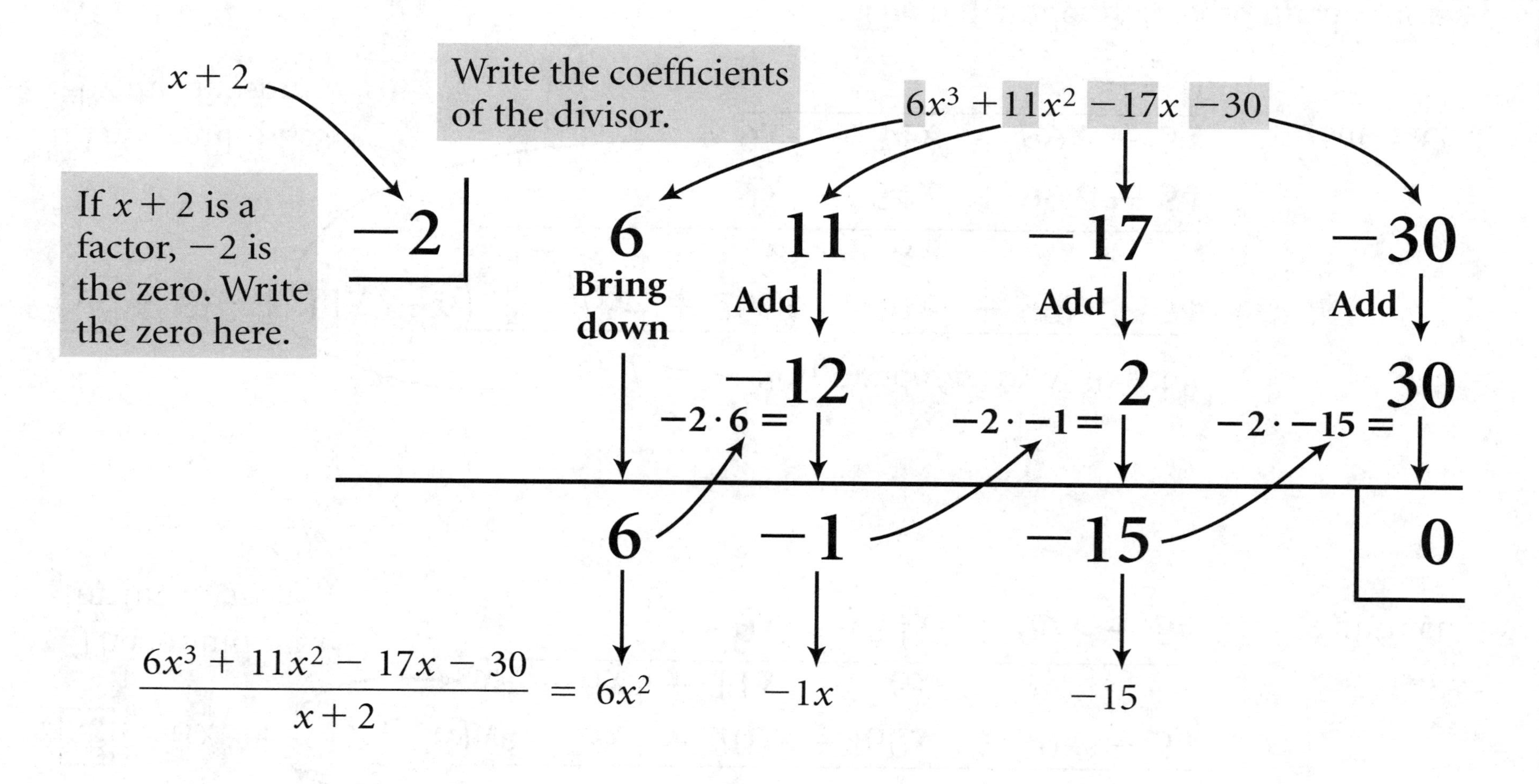

x + 2
Write the coefficients of the divisor.
6x³ + 11x² − 17x − 30
If x + 2 is a factor, −2 is the zero. Write the zero here.
−2
6
11
−17
−30
Bring down
Add
Add
Add
−12
2
30
−2 · 6 =
−2 · −1 =
−2 · −15 =
6
−1
−15
0
(6x³ + 11x² − 17x − 30) / (x + 2) = 6x² −1x −15

Review Data

-3	-56
-2	-12
-1	-12
0	25
1	48
2	99
3	196

Bucket Race

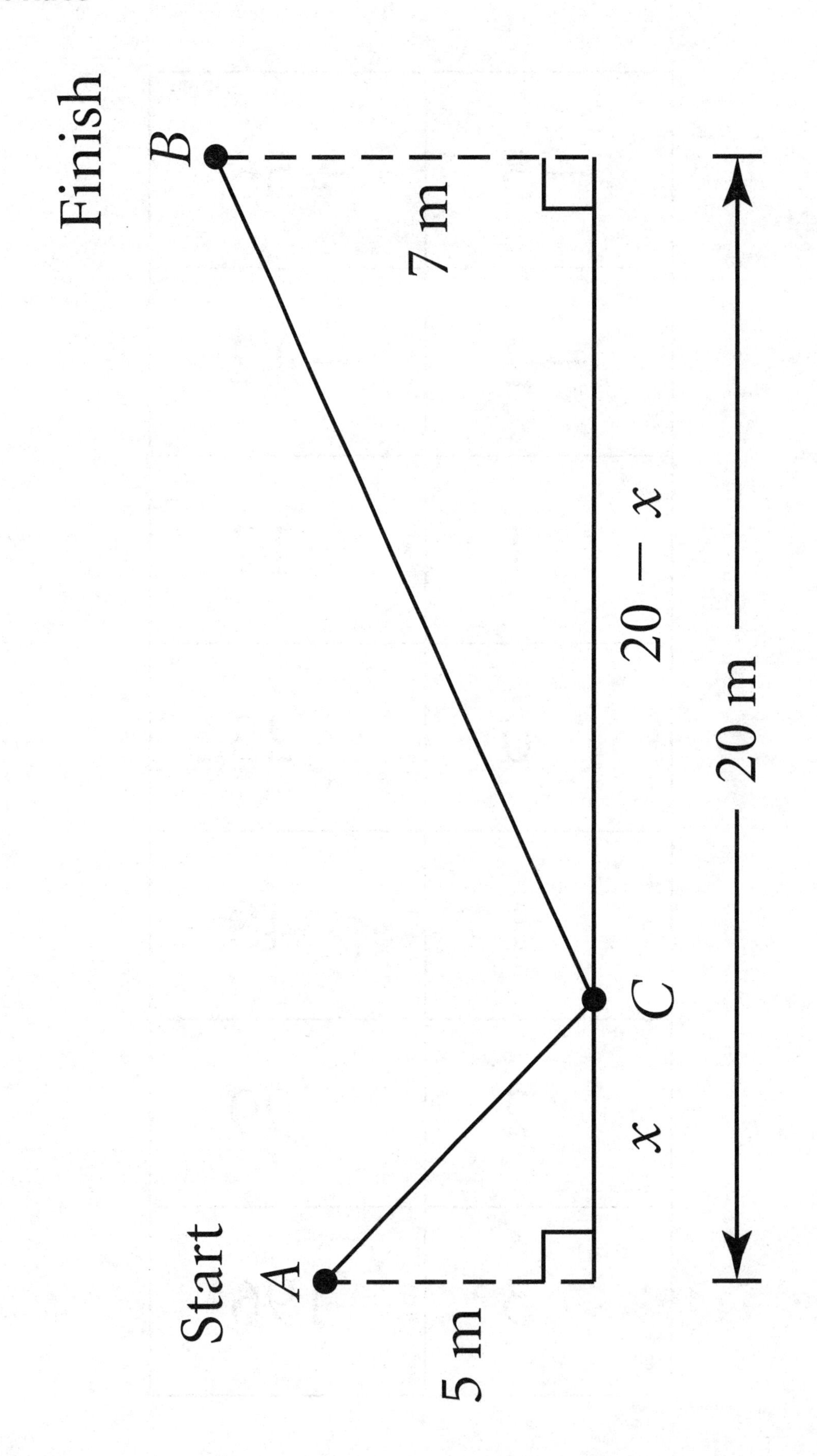

Oil Rig Rescue

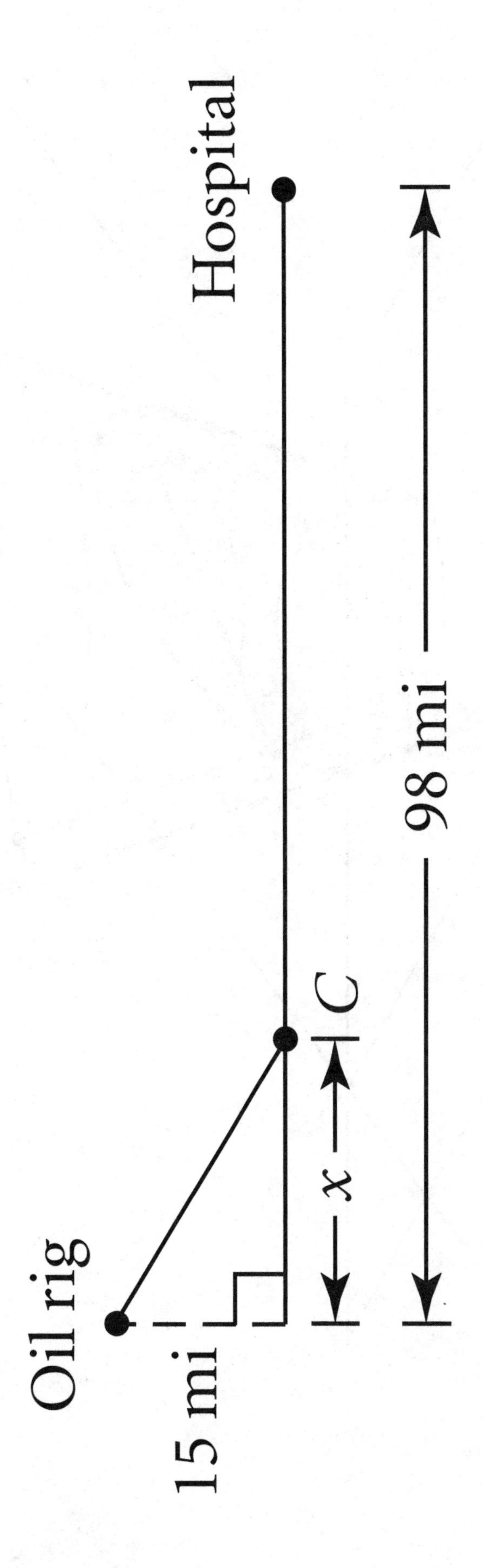

Ellipse in a Cone

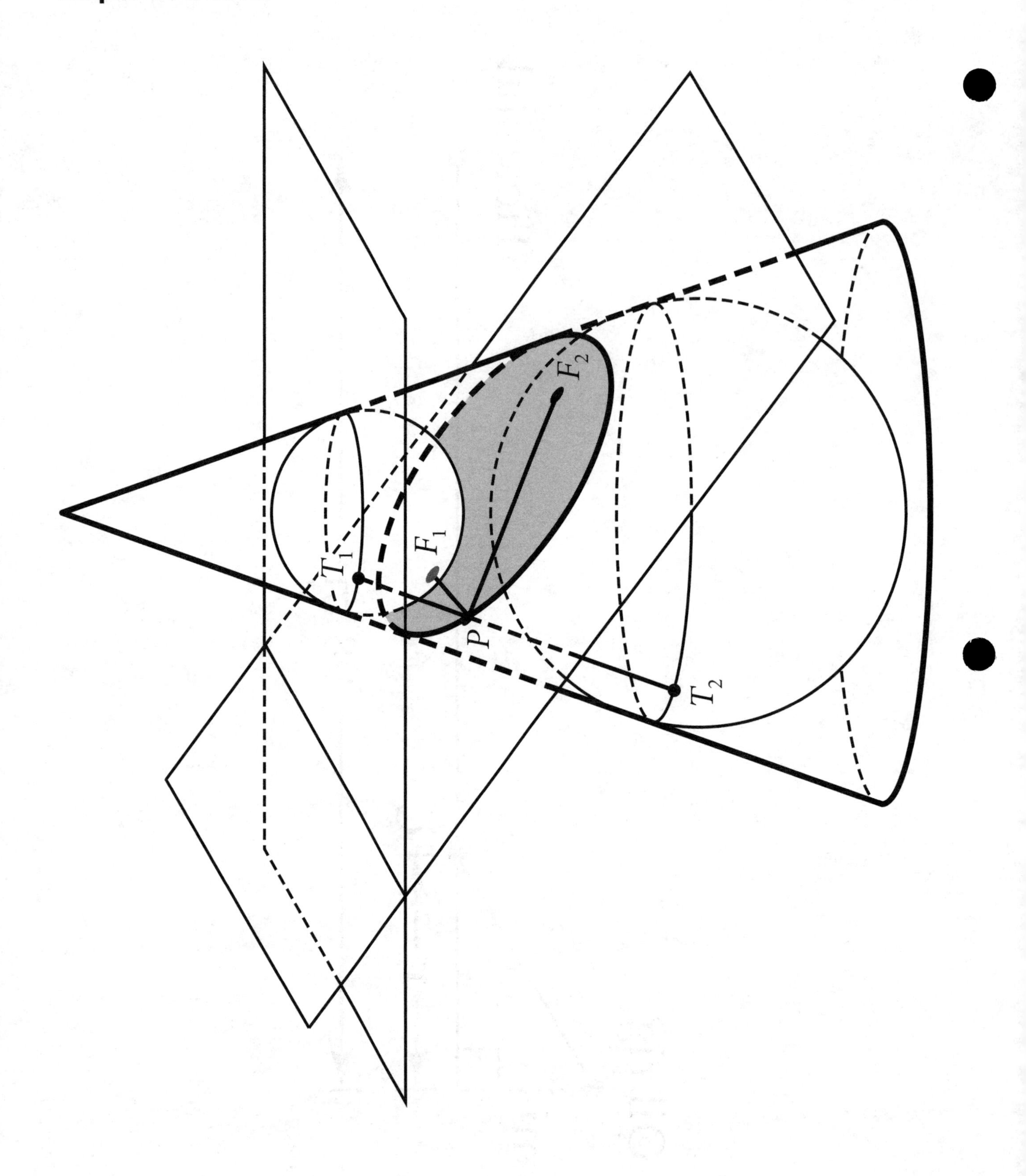

Definitions of Circle and Ellipse

Circle

A **circle** is a locus of points P in a plane that are a constant distance, r, from a fixed point, C. Symbolically, $PC = r$. The fixed point is called the **center** and the constant distance is called the **radius.**

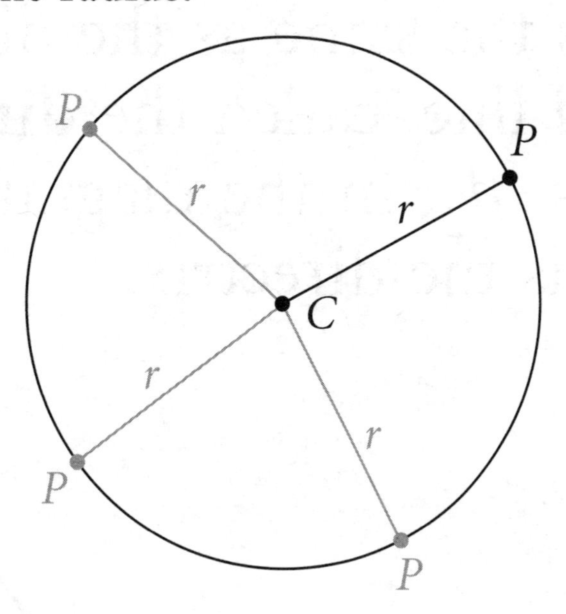

Ellipse

An **ellipse** is a locus of points in a plane, the sum of whose distances from two fixed points is always a constant. In the diagram, the two fixed points, or **foci** (the plural of focus), are labeled F_1 and F_2. For all points on the ellipse, the distances d_1 and d_2 sum to the same value.

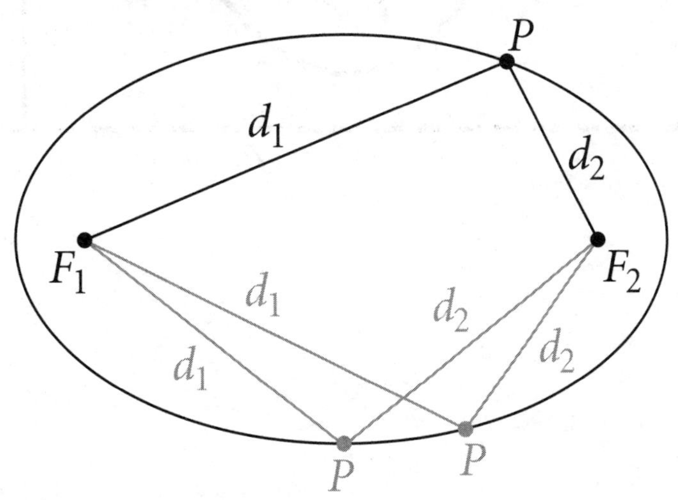

A **parabola** is a locus of points in a plane whose distance from a fixed point, called the **focus,** is the same as the distance from a fixed line, called the **directrix.** That is, $d_1 = d_2$. In the diagram, F is the focus and l is the directrix.

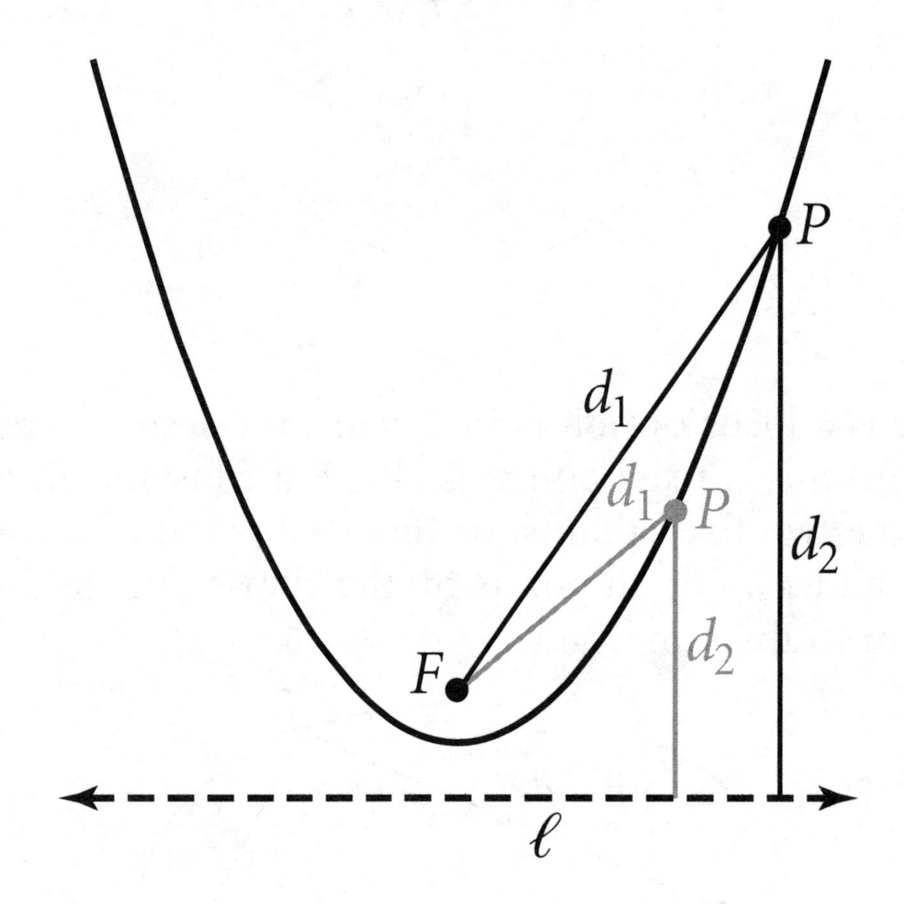

Hyperbola in a Cone

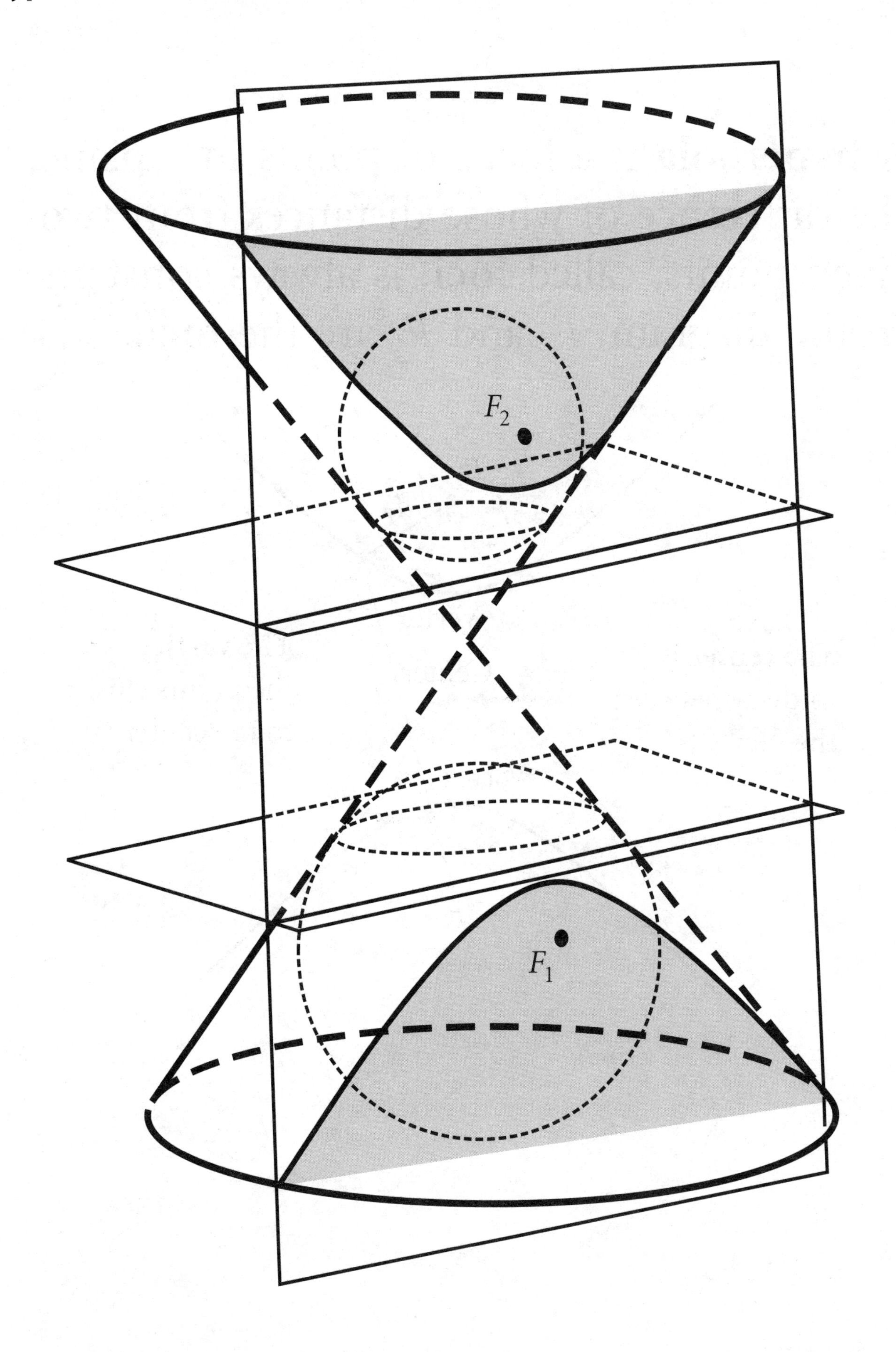

A **hyperbola** is a locus of points in a plane, the difference of whose distances from two fixed points, called **foci,** is always constant. In the diagram, F_1 and F_2 are the **foci.**

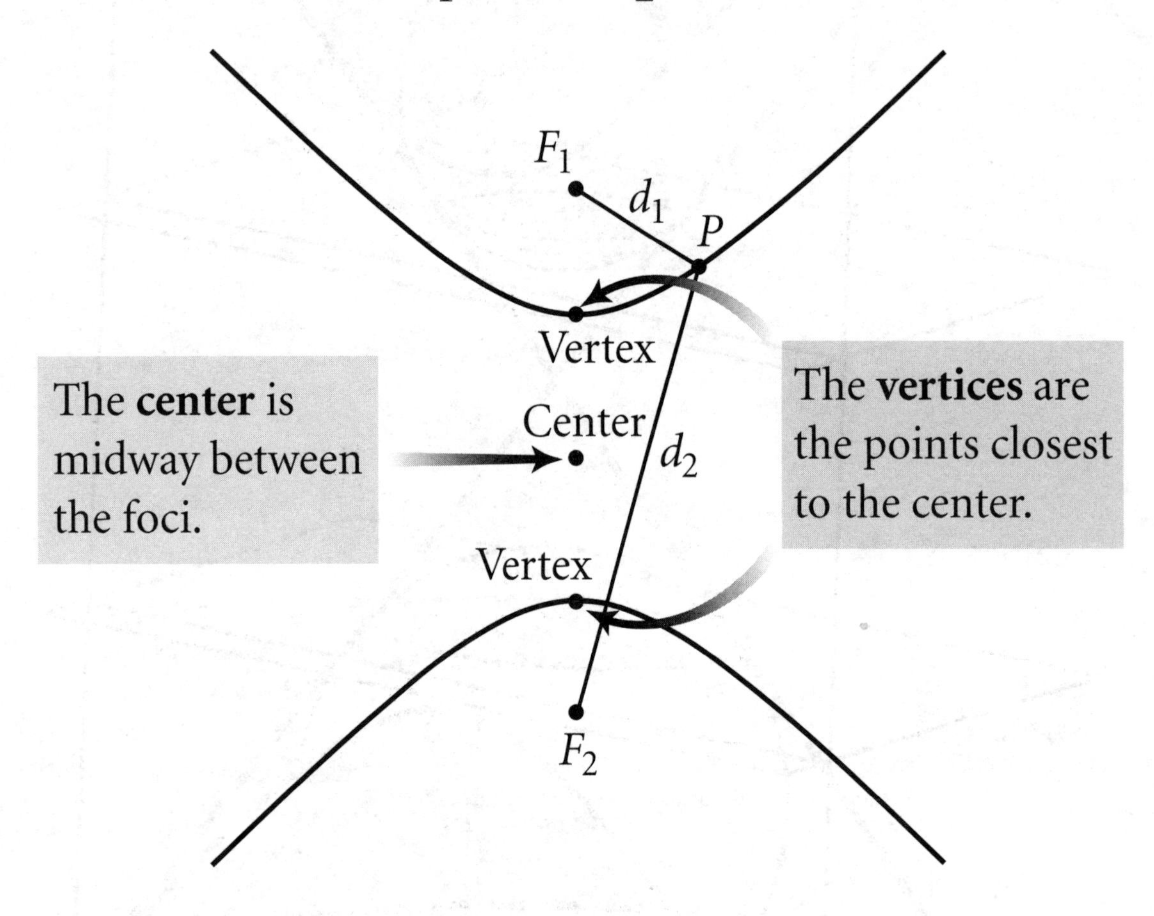

Passing By Sample Data

Time (s)	Distance (m)
0	5.000
0.2	4.720
0.4	4.381
0.6	4.016
0.8	3.688
1.0	3.558
1.2	3.302
1.4	3.078
1.6	2.709
1.8	2.410
2.0	2.249
2.2	1.969
2.4	1.770
2.6	1.605
2.8	1.525
3.0	1.291
3.2	1.022
3.4	0.901
3.6	0.882

Time (s)	Distance (m)
3.8	0.926
4.0	0.966
4.2	1.056
4.4	1.240
4.6	1.387
4.8	1.568
5.0	1.673
5.2	1.978
5.4	2.251
5.6	2.478
5.8	2.748
6.0	3.099
6.2	3.284
6.4	3.533
6.6	3.820
6.8	4.116
7.0	4.379
7.2	4.695
7.4	4.955

Rational Function Graphs

a.

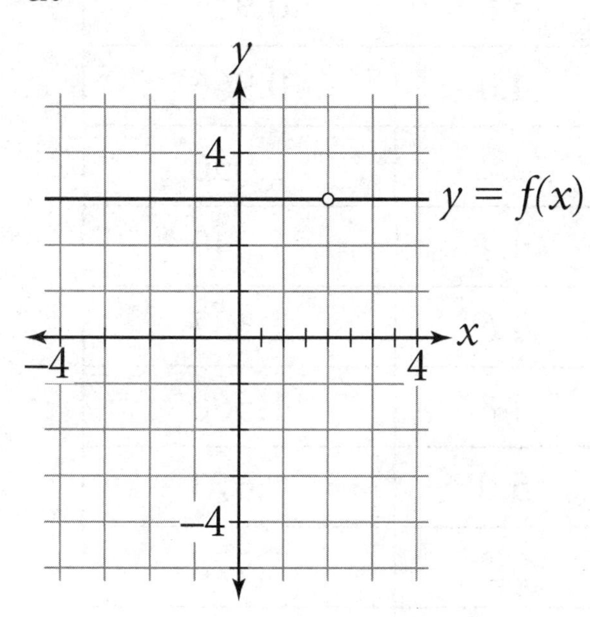

b.

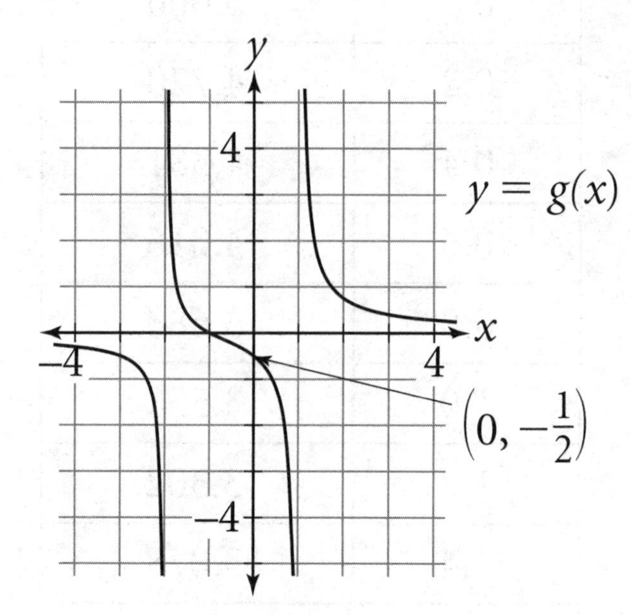

c.

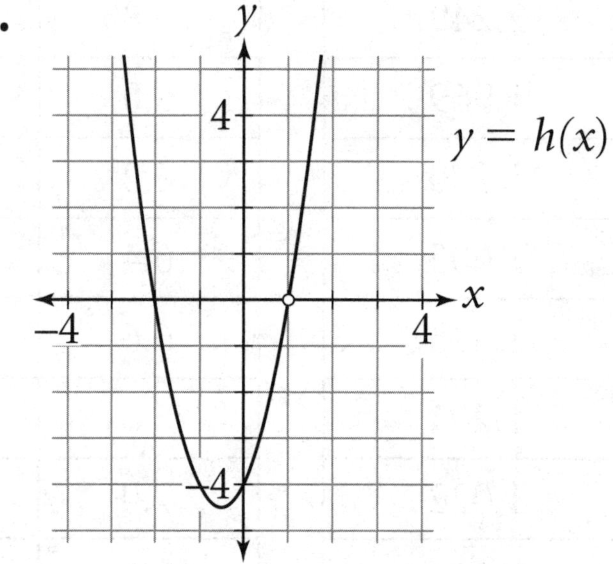

Exercise 9

a.

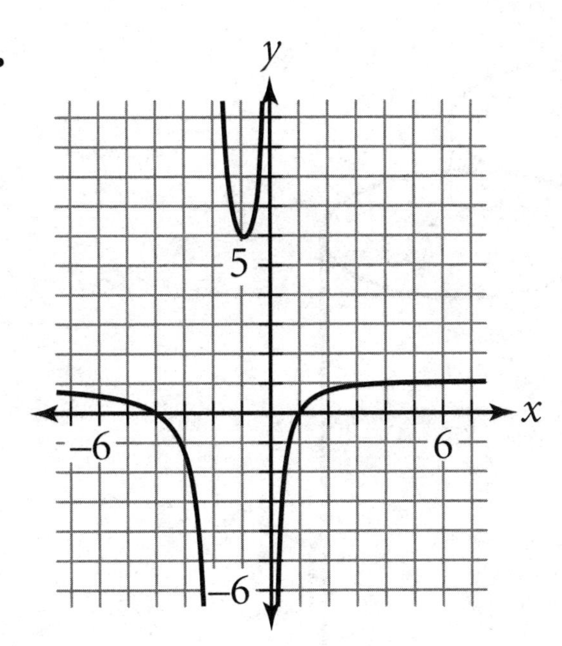

b.

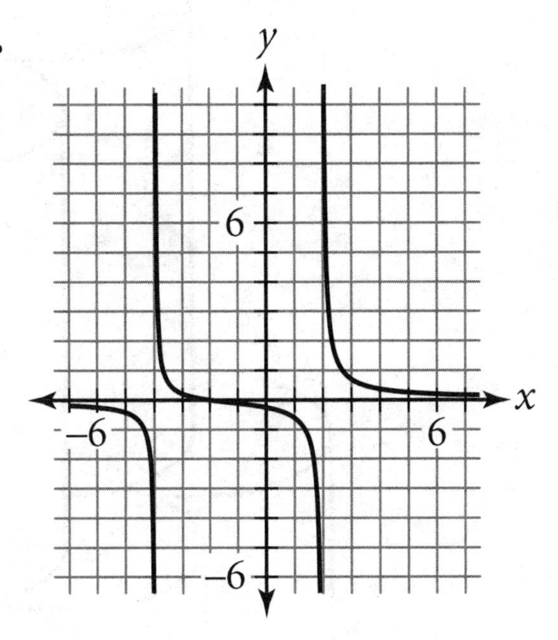

c.

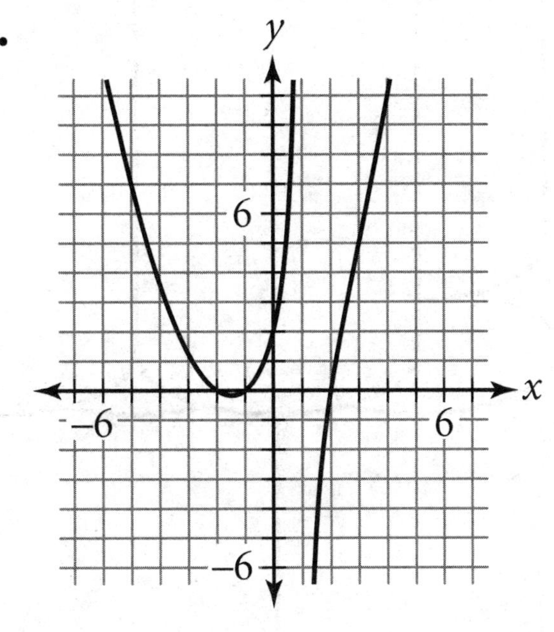

Soup Cans

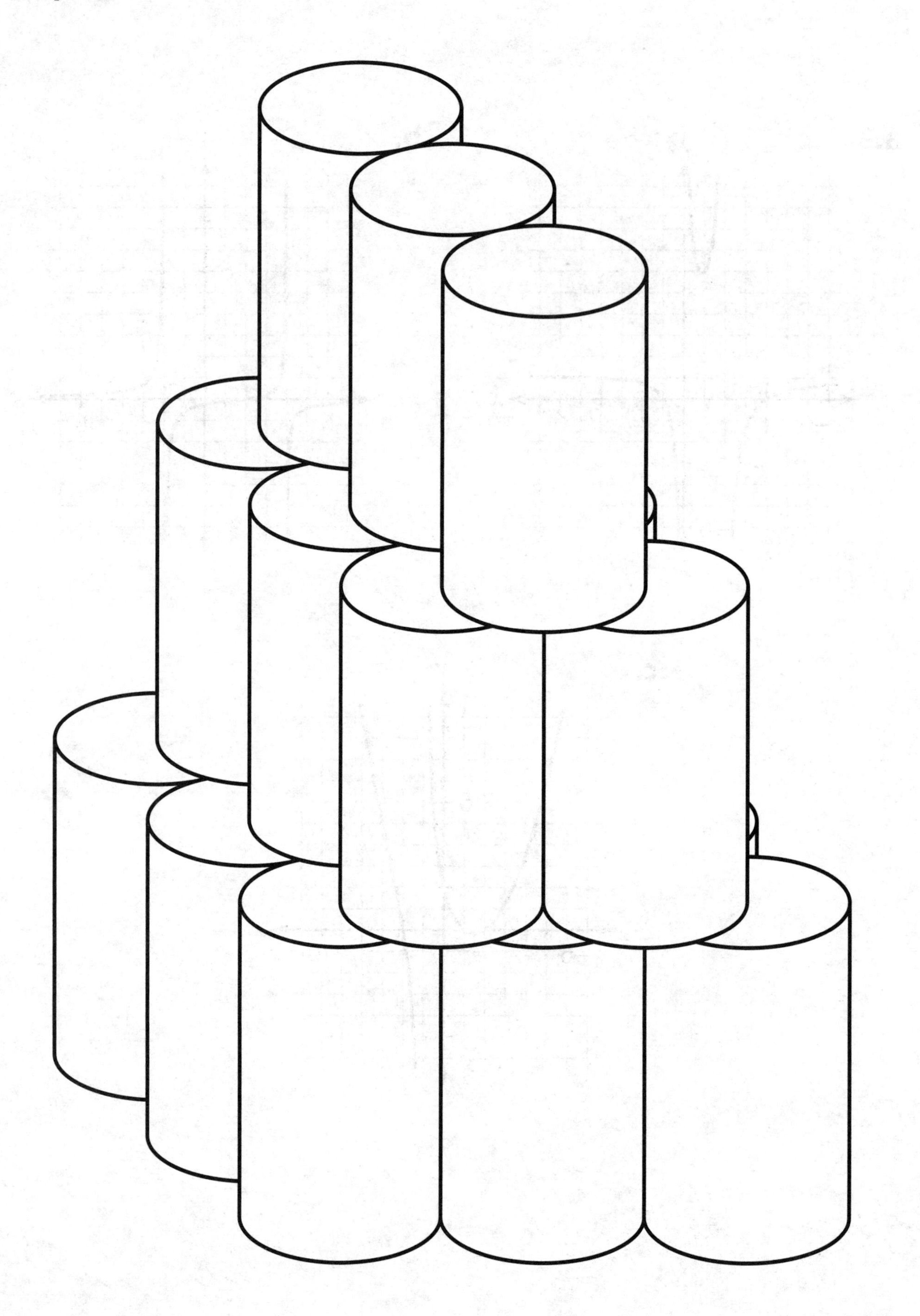

Arithmetic Series Formula

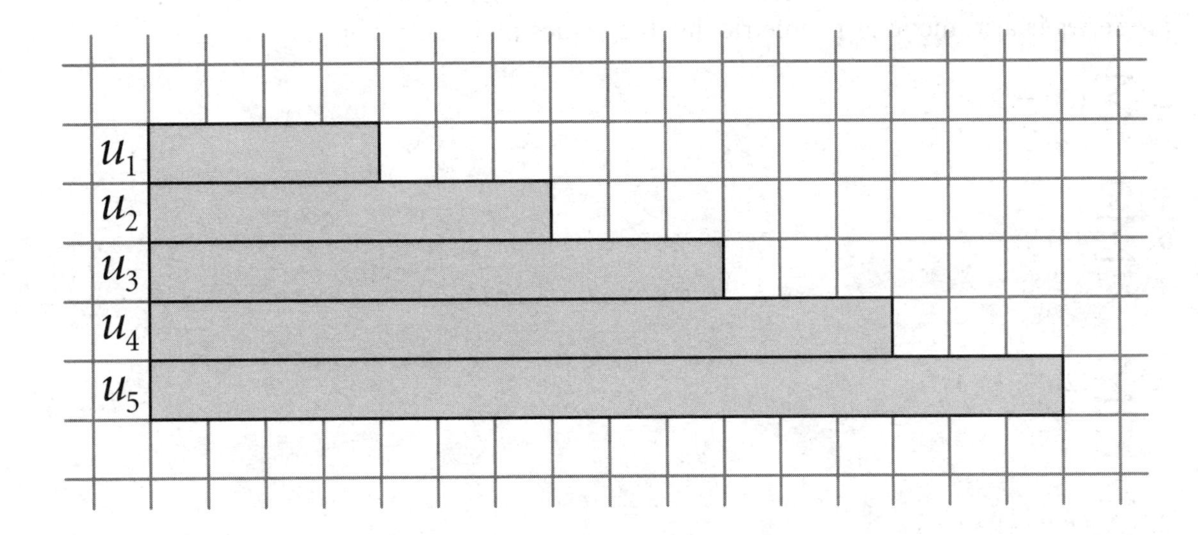

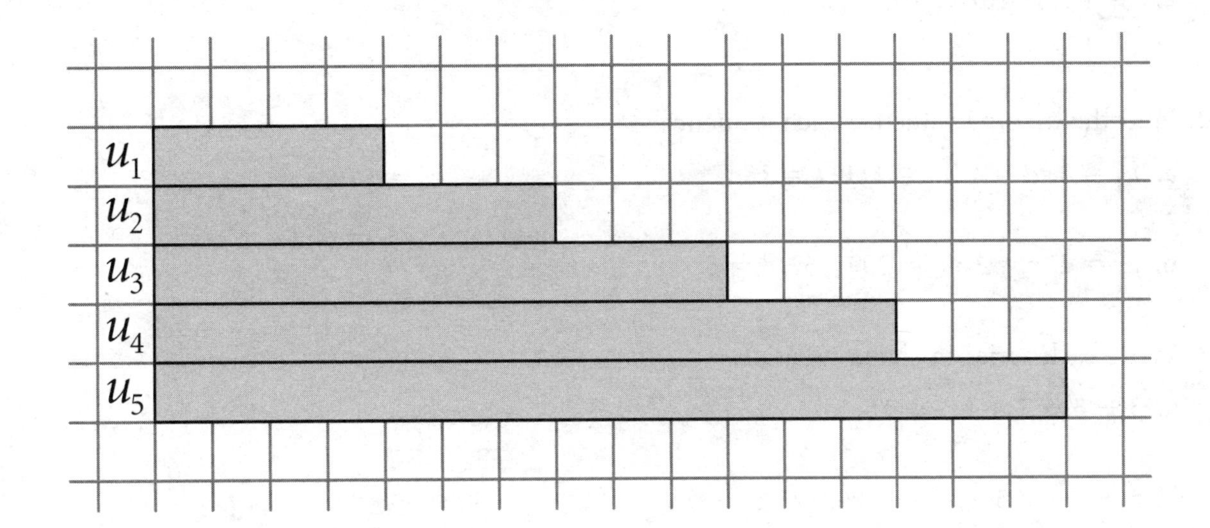

Sigma Notation

Name ______________________________ Period __________ Date ______________

1. Write each expression as a sum of terms, then calculate the sum. If the series is arithmetic or geometric, list the values of u_1 and d or r.

a. $\sum_{n=1}^{8}(3n + 2)$

b. $\sum_{n=1}^{5}4(0.8)^{n-1}$

c. $\sum_{n=1}^{5}(2n) \cdot (-1)^n$

d. $\sum_{n=1}^{4}(n + 1) \cdot (-1)^{n+1}$

e. $\sum_{n=1}^{10}(n^2 - 10n)$

2. Find the missing value for each sequence.

a. $u_1 = 3,\ d = 4,\ u_k = 111,\ k = ?$

b. $u_1 = 2,\ r = 2,\ u_k = 2097152,\ k = ?$

3. Write each series in sigma notation.

a. $3 + 5 + 7 + \cdots + 27$

b. $2 + 1 + 0.5 + \cdots + 0.03125$

c. $5 - 15 + 25 - \cdots + 225$

Flip a Coin

Name ______________________ Period __________ Date ______________

Sequence A (imagined sequence)										
Sequence B (actual sequence)										

Longest string	Sequence A	Sequence B	2nd-longest string	Sequence A	Sequence B
1			1		
2			2		
3			3		
4			4		
5 or more			5 or more		

Number of H's	Sequence A	Sequence B
0		
1		
2		
3		
4		
5		
6		
7		
8		
9		
10		

Exercise 8

Simulation number	1's	2's	3's	4's	5's	6's	Ratio of 3's	Cumulative ratio of 3's
1								$\frac{\quad}{100} = \_\_\_$
2								$\frac{\quad}{100} = \_\_\_$
3								$\frac{\quad}{100} = \_\_\_$
4								
5								
6								
7								
8								
9								
10								
11								
12								

The Coin Toss Problem

	Coin diameter (mm)	20 mm grid	30 mm grid	40 mm grid	40 mm grid with 5 mm borders
Penny					
Nickel					
Dime					
Quarter					

20 mm Grid Paper

30 mm Grid Paper

40 mm Grid Paper

40 mm Grid Paper with 5 mm Borders

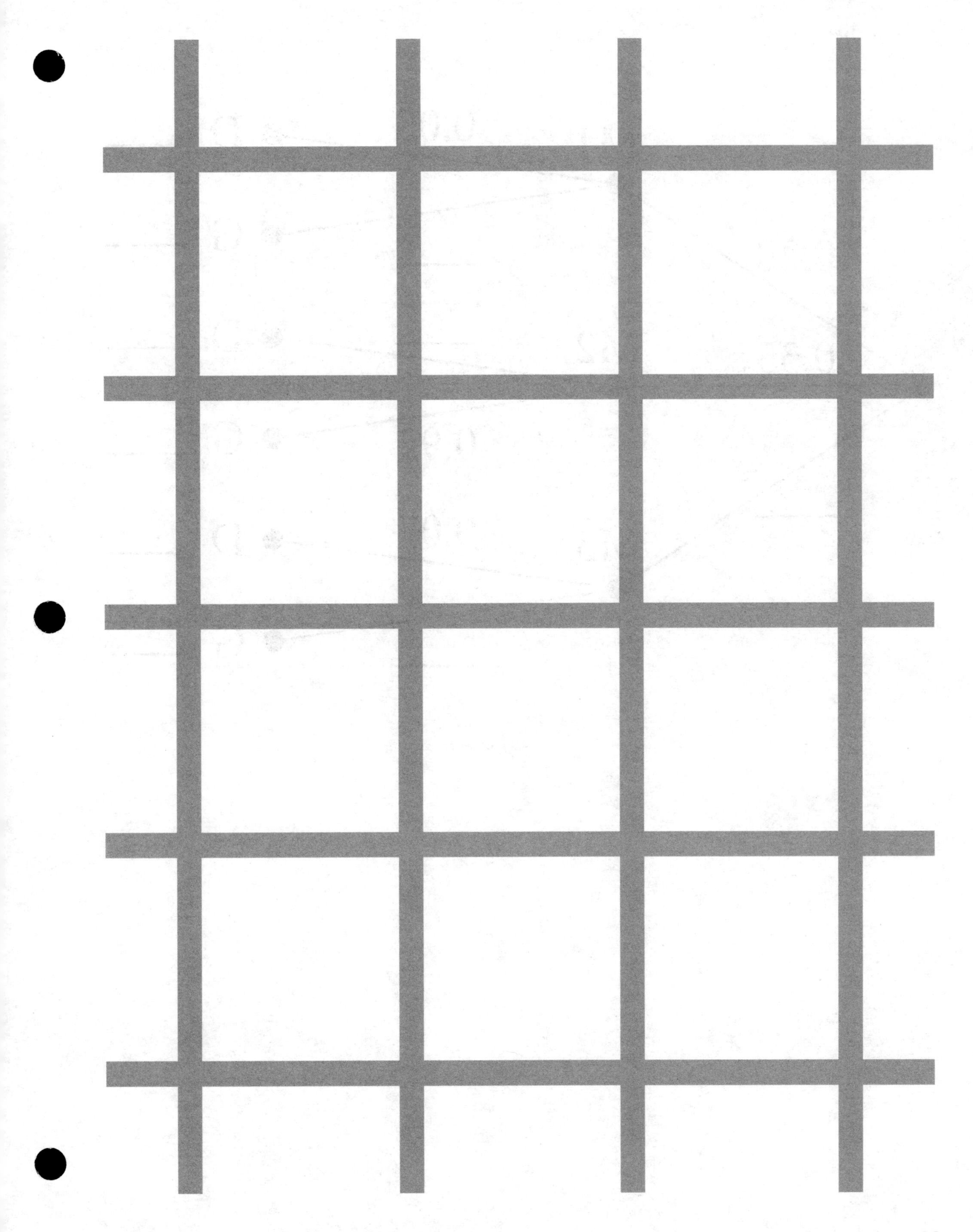

Exercise 12

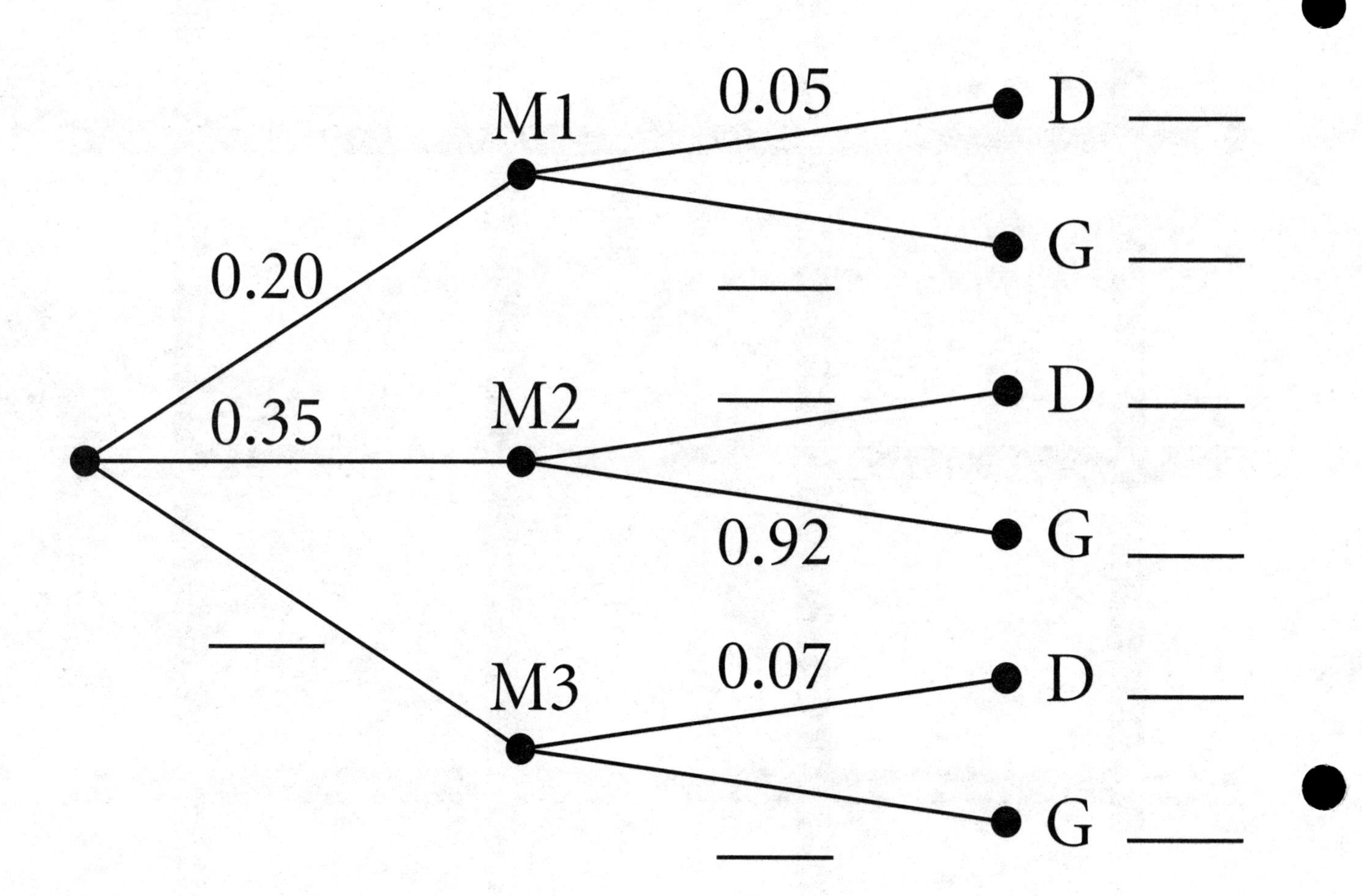

Venn Diagram

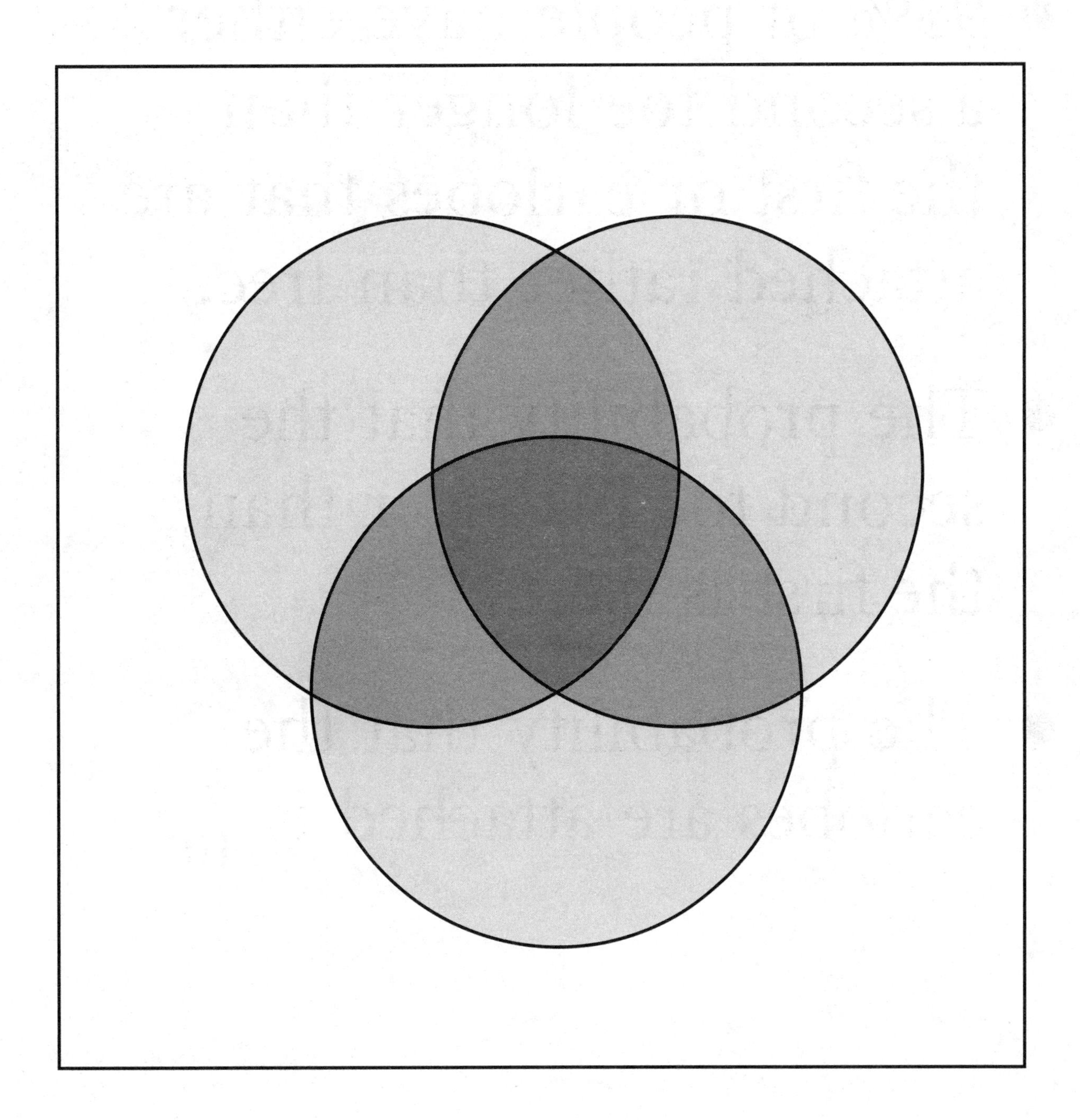

- 95% of people have either a second toe longer than the first or earlobes that are attached rather than free.
- The probability that the second toe is longer than the first is 0.6.
- The probability that the earlobes are attached is $\frac{7}{10}$.

Pascal's Triangle

1

1 1

1 2 1

1 3 3 1

1 4 6 4 1

1 5 10 10 5 1

1 6 15 20 15 6 1

1 7 21 35 35 21 7 1

1 8 28 56 70 56 28 8 1

1 9 36 84 126 126 84 36 9 1

1 10 45 120 210 252 210 120 45 10 1

Random-Number Generator

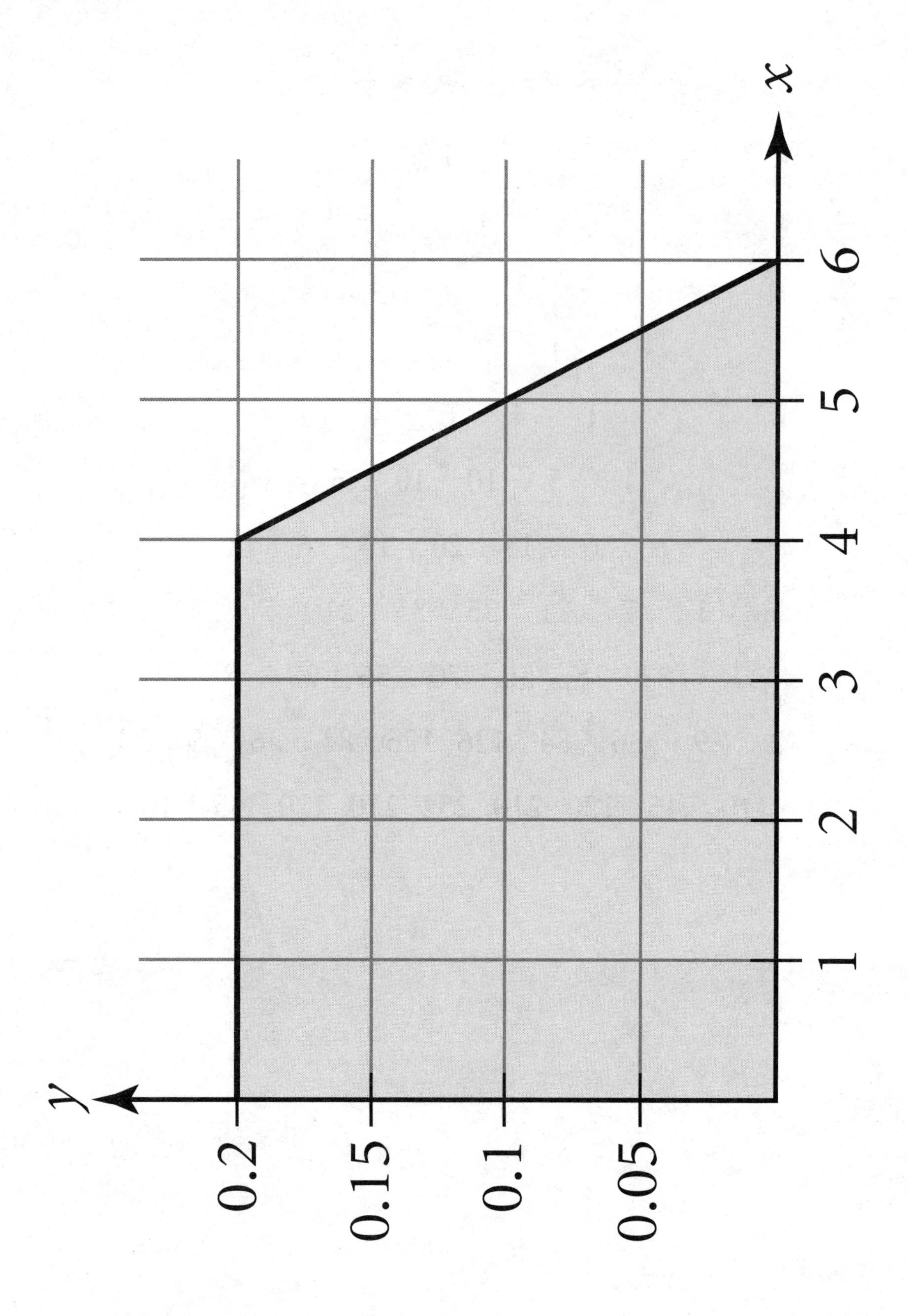

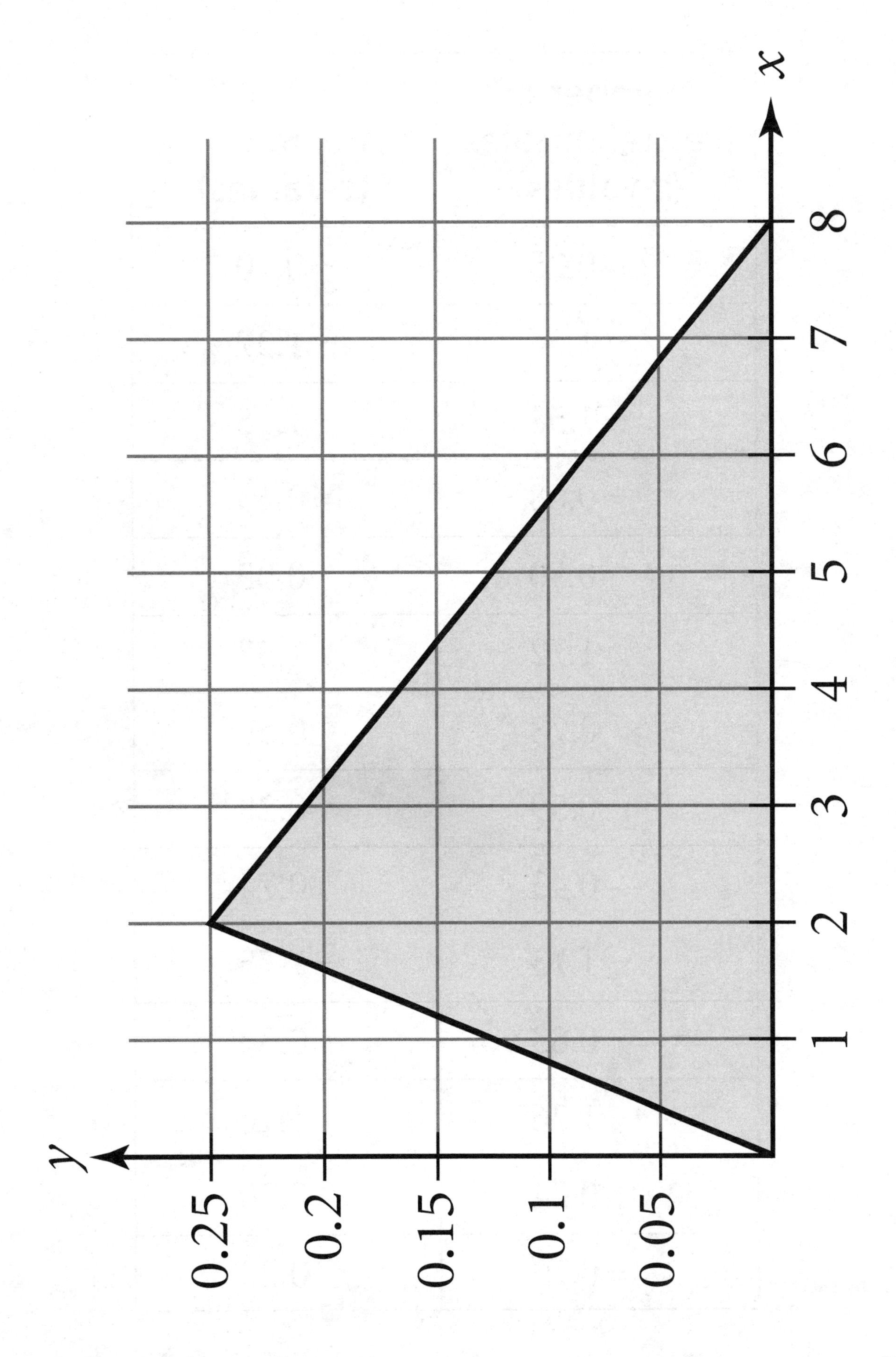
x
8
7
6
5
4
3
2
1
y
0.25
0.2
0.15
0.1
0.05

Sunburn and Ice Cream

Number of ice cream sales (z-values)	Visits to nurse (z-values)
0.25	−0.10
−1.36	−1.37
−1.58	−1.78
−0.08	−0.13
0.50	0.05
1.72	1.57
0.78	0.59
0.59	0.28
0.43	0.74
−1.13	−1.75
−0.83	0.36
1.26	1.10
0.24	0.20
−0.79	0.23

Looking for Connections Sample Data

Student	Question 1	Question 2	Question 3	Question 4	Question 5
1	0	120	60	100	3
2	20	120	90	200	4
3	80	65	80	140	6
4	55	20	220	260	5
5	10	20	155	200	4
6	15	0	145	200	4
7	90	10	80	150	6
8	215	10	0	60	6
9	100	0	140	150	6
10	60	30	120	105	5
11	65	0	120	150	6
12	10	60	300	360	4
13	120	0	0	45	6
14	30	45	285	90	4
15	40	60	150	190	4
16	0	85	150	75	3
17	0	180	30	30	4
18	80	0	0	0	6
19	90	20	0	0	6
20	45	10	180	285	5
21	10	120	0	90	4
22	40	30	100	115	5
23	0	0	360	60	5
24	30	50	45	90	5
25	60	20	30	90	6
26	45	20	30	45	5
27	20	105	20	60	4
28	0	90	0	120	4
29	50	40	0	90	5
30	40	10	0	45	4

Spin Time Sample Data

Coin	Spin time (s)
Quarter	10.69
Quarter	12.26
Quarter	13.56
Quarter	9.71
Dime	8.38
Dime	7.47
Dime	8.87
Dime	7.37
Nickel	15.48
Nickel	13.88
Nickel	12.19
Nickel	12.54
Cent	9.6
Cent	11.39
Cent	8.25
Cent	6.78

Age and Income

Age (yr)	Income ($)
31	6,000
43	0
48	35,000
5	0
41	60,000
37	65,970
21	20
50	3,000
35	68,400
48	22,330
10	0
74	36,636
32	44,000
54	8,500
54	27,000
42	50,125
16	600

Age (yr)	Income ($)
45	31,300
38	34,500
77	18,000
46	63,058
2	0
67	65,568
33	50,592
54	80,000
16	0
14	0
44	106,000
8	0
56	36,868
42	20,000
39	6,500
61	35,000
19	750

Age (yr)	Income ($)
22	2,400
40	17,000
27	10,090
22	15,000
38	72,000
39	51,000
30	0
36	1,500
4	0
22	1,561
37	52,523
23	6,000
40	15,462
31	24,634
3	0
42	87,500

Steep Steps

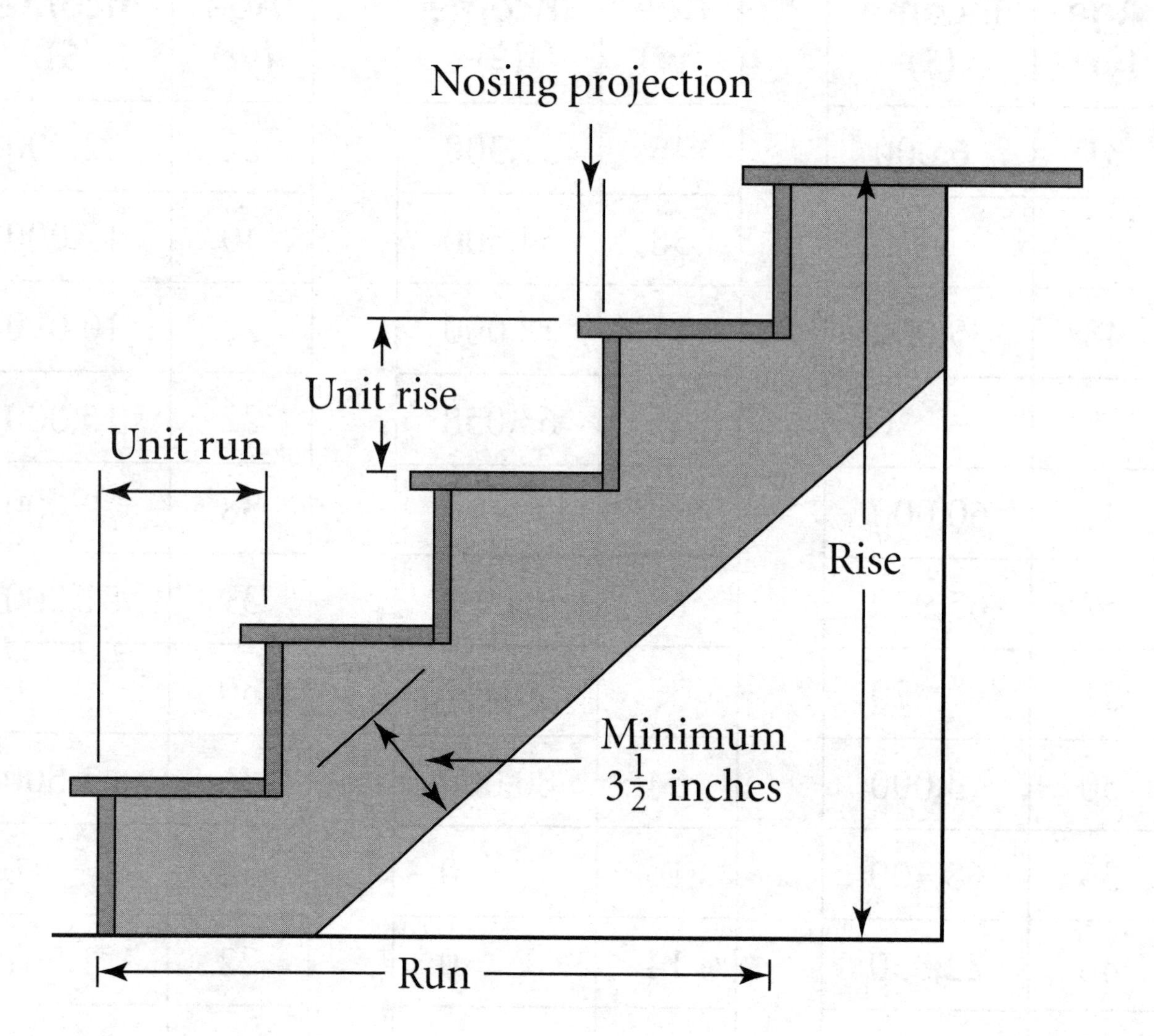
Nosing projection
Unit rise
Unit run
Rise
Minimum
$3\frac{1}{2}$ inches
Run

Trigonometric Ratios

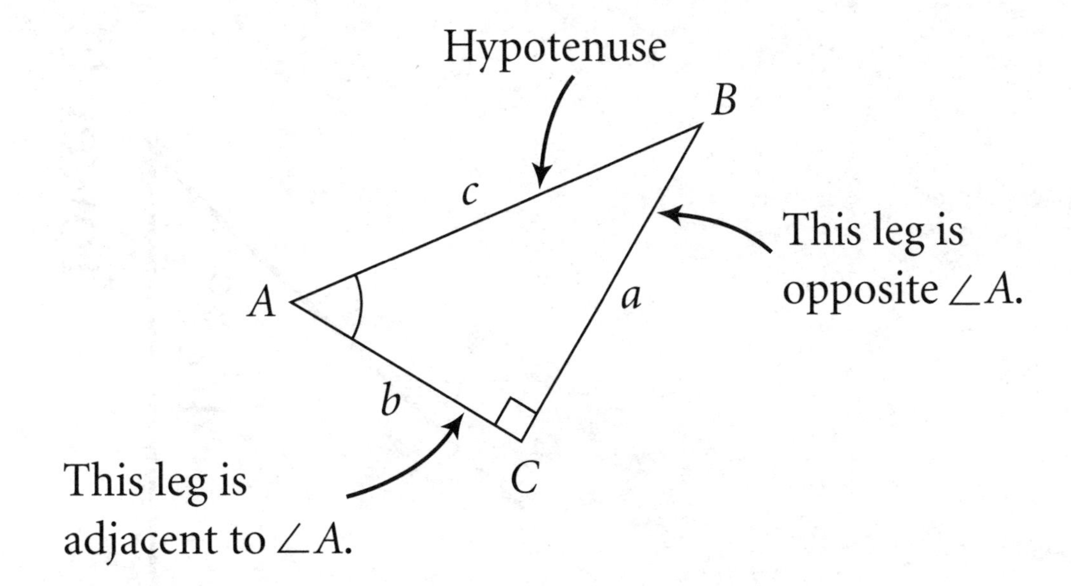

For any acute angle A in a right triangle, the **sine** of $\angle A$ is the ratio of the length of the leg opposite $\angle A$ to the length of the hypotenuse.

$$\sin A = \frac{\text{opposite leg}}{\text{hypotenuse}} = \frac{a}{c}$$

The **cosine** of $\angle A$ is the ratio of the length of the leg adjacent to $\angle A$ to the length of the hypotenuse.

$$\cos A = \frac{\text{adjacent leg}}{\text{hypotenuse}} = \frac{b}{c}$$

The **tangent** of $\angle A$ is the ratio of the length of the opposite leg to the length of the adjacent leg.

$$\tan A = \frac{\text{opposite leg}}{\text{adjacent leg}} = \frac{a}{b}$$

Towers

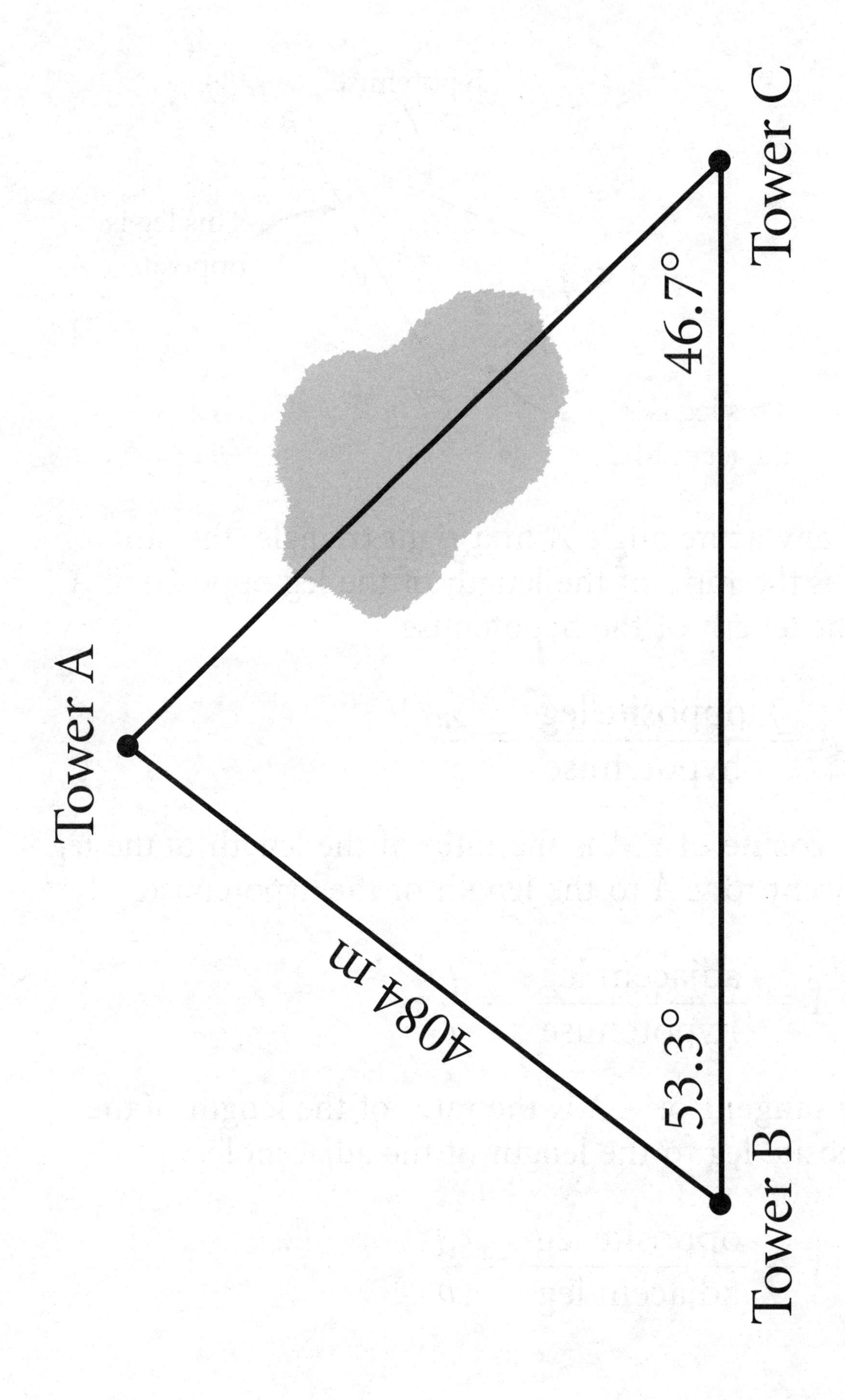

Parametric Walk Sample Data

t	x	y
0.1	1.78	1.95
0.6	1.71	2.02
1.1	1.62	2.11
1.6	1.50	2.12
2.1	1.43	2.21
2.6	1.36	2.25
3.1	1.26	2.32
3.6	1.19	2.38
4.1	1.10	2.39
4.6	1.00	2.47
5.0	0.95	2.50

Defining Circular Functions

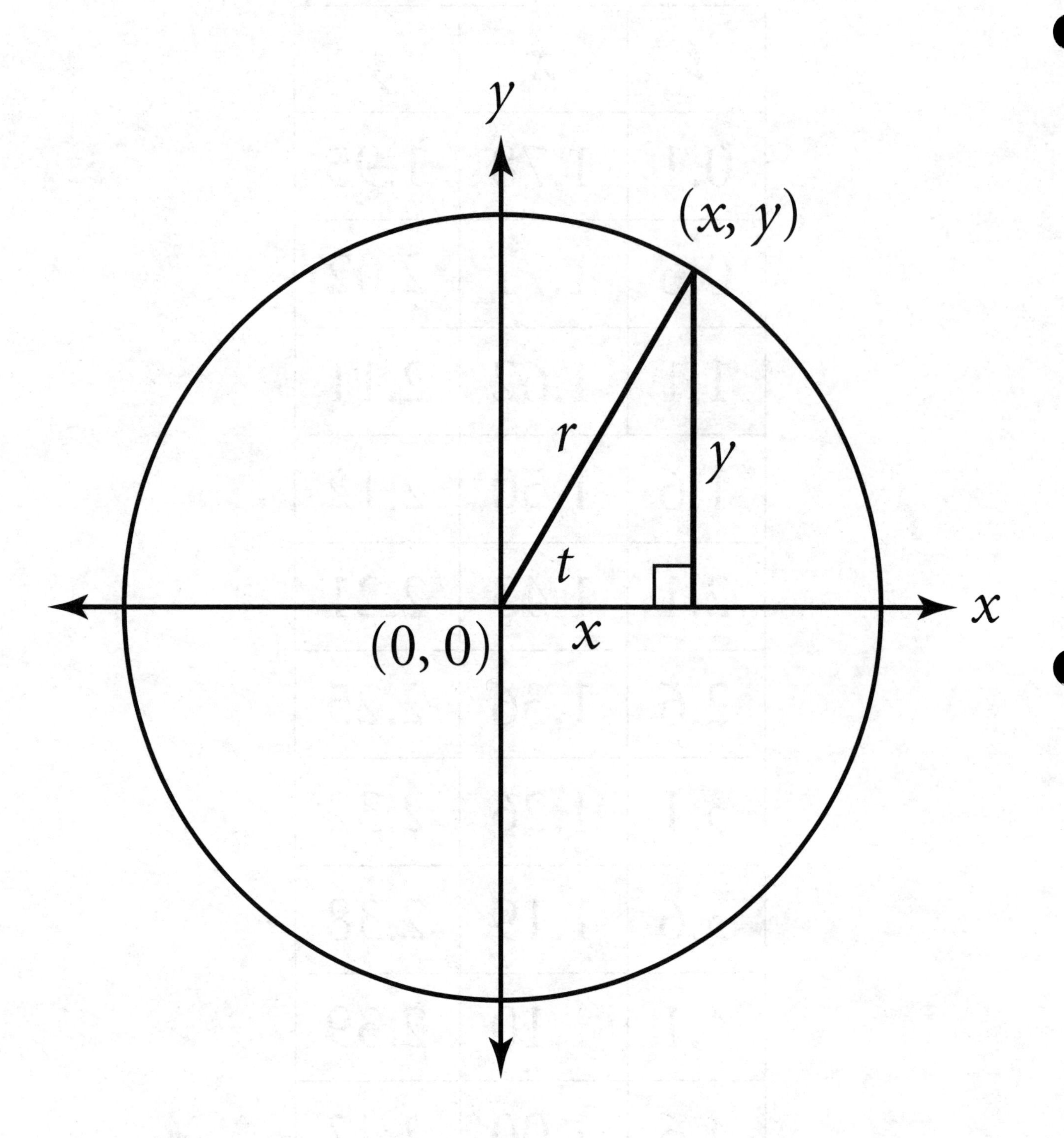

A Radian Protractor

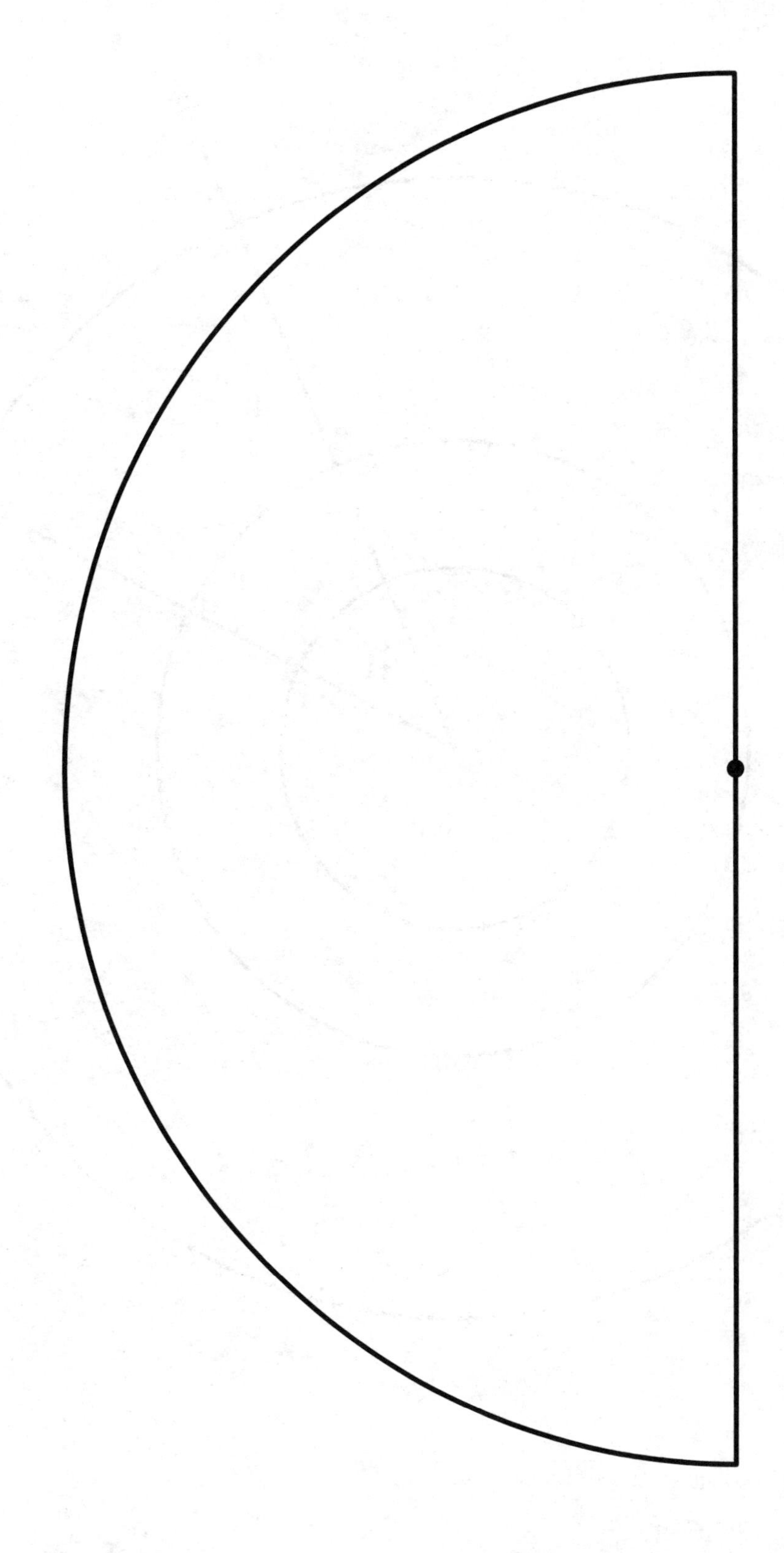

Arcs and Radii

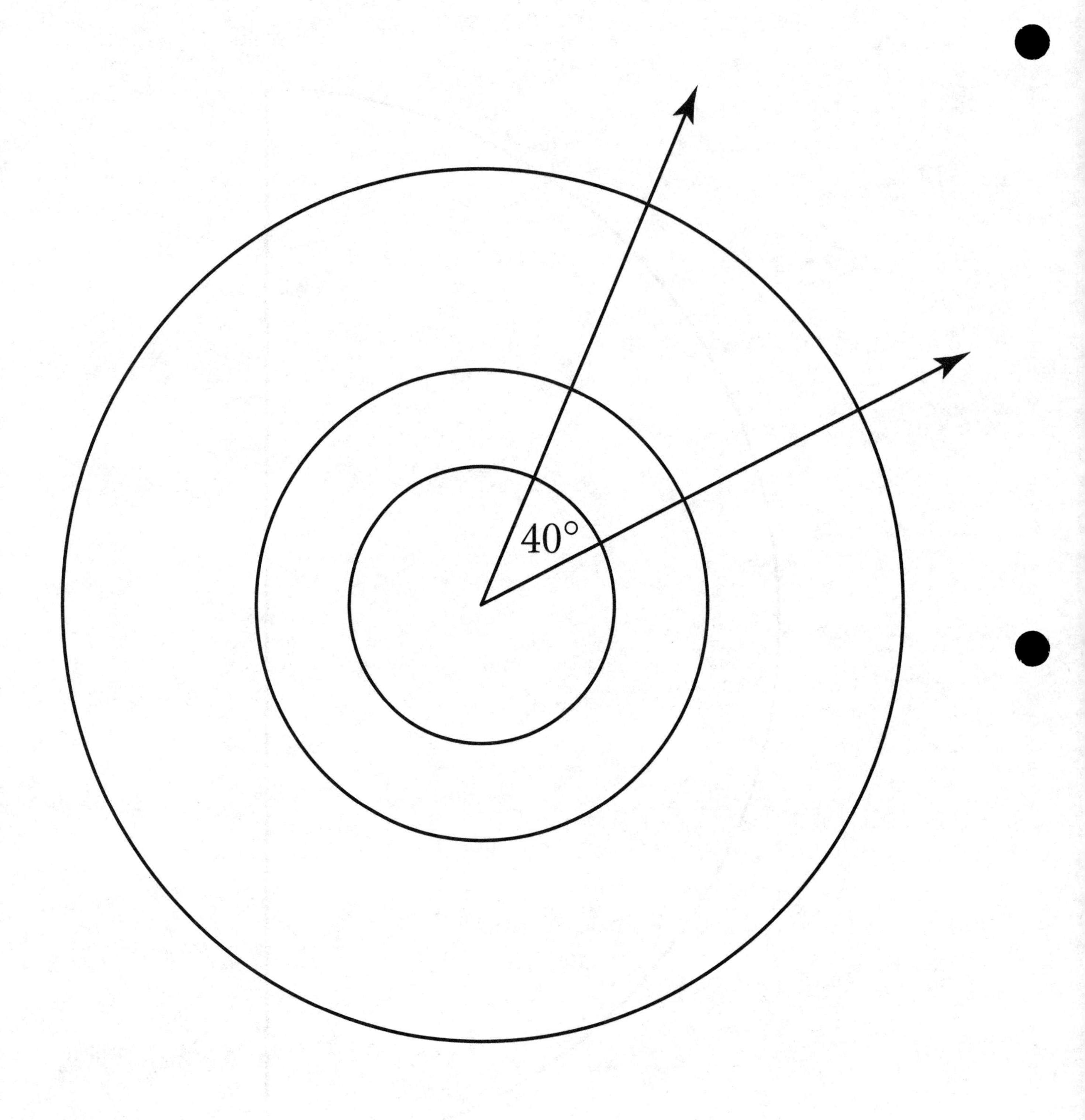

Exercise 3

Name ______________________ Period __________ Date ______________

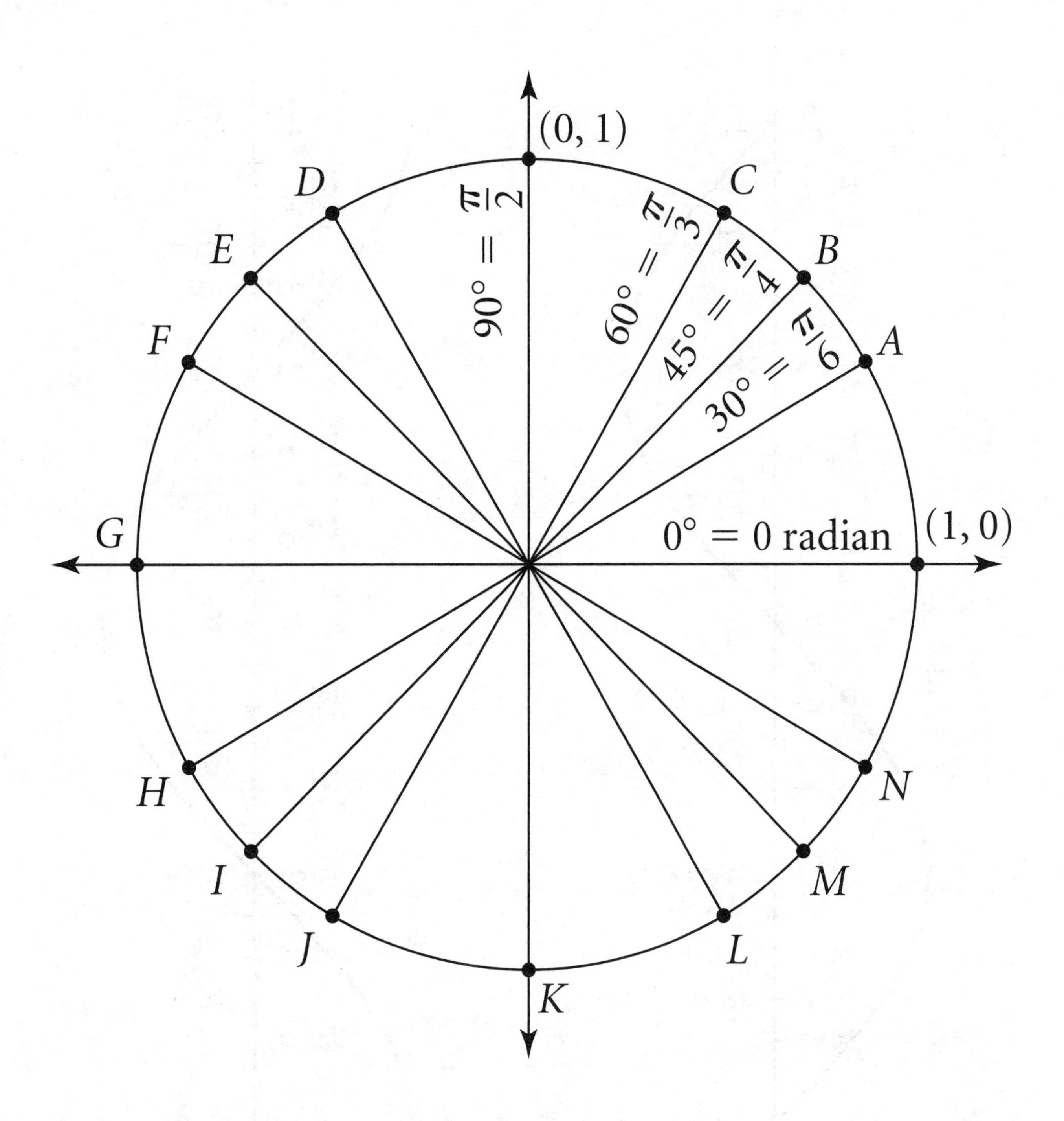

Exercises 7 and 8 Solutions

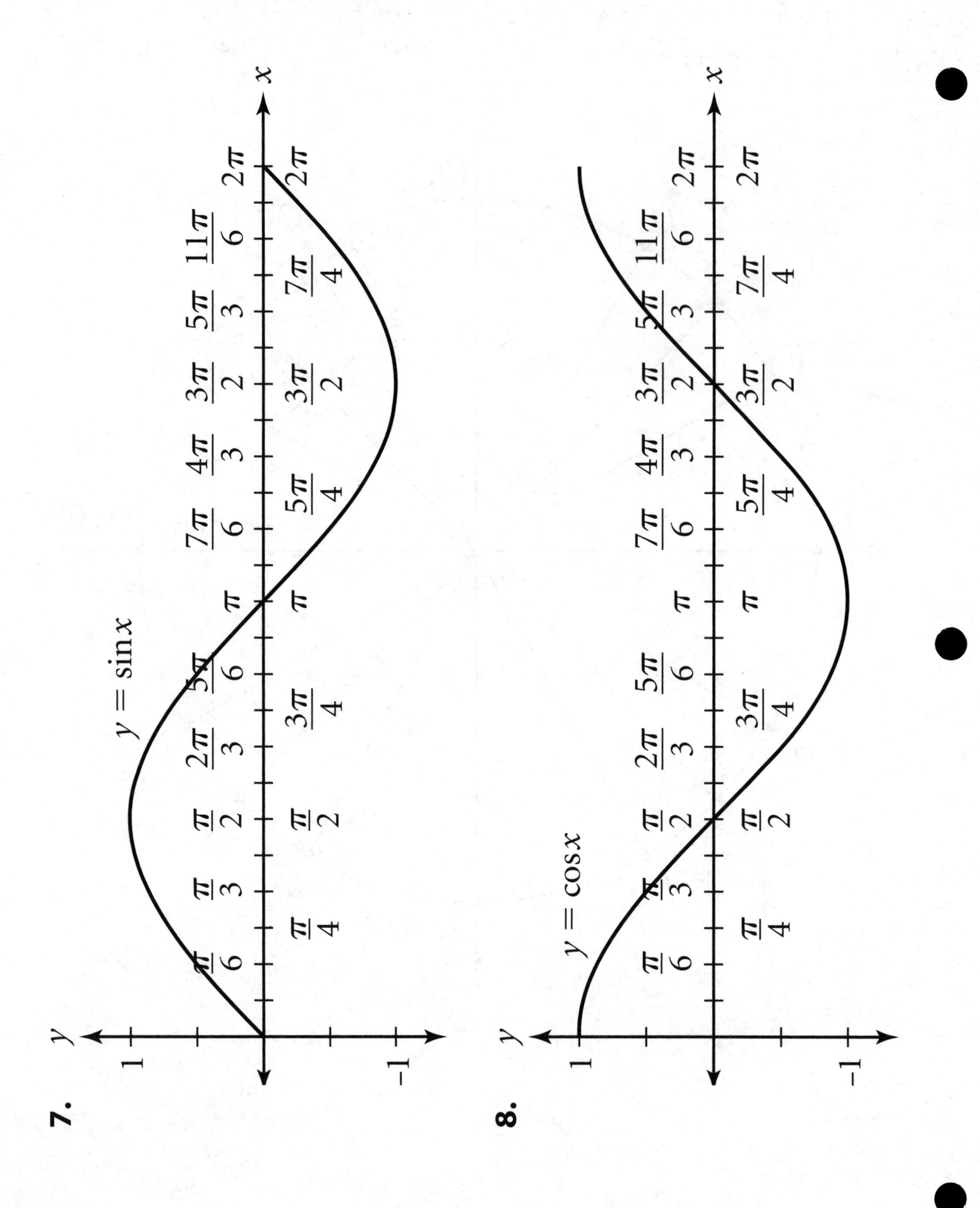

Pendulum II Sample Data

Time (s)	Distance (m)
0.05	0.900
0.10	0.903
0.15	0.899
0.20	0.888
0.25	0.873
0.30	0.855
0.35	0.836
0.40	0.814
0.45	0.791
0.50	0.768
0.55	0.749
0.60	0.730
0.65	0.714
0.70	0.704
0.75	0.698
0.80	0.697
0.85	0.702
0.90	0.712
0.95	0.727
1.00	0.745

Time (s)	Distance (m)
1.05	0.766
1.10	0.786
1.15	0.809
1.20	0.830
1.25	0.850
1.30	0.867
1.35	0.881
1.40	0.891
1.45	0.894
1.50	0.894
1.55	0.889
1.60	0.879
1.65	0.863
1.70	0.844
1.75	0.824
1.80	0.802
1.85	0.779
1.90	0.760
1.95	0.740
2.00	0.724

Coordinate Axes for Periodic Functions

Name ______________________________ Period ____________ Date ______________

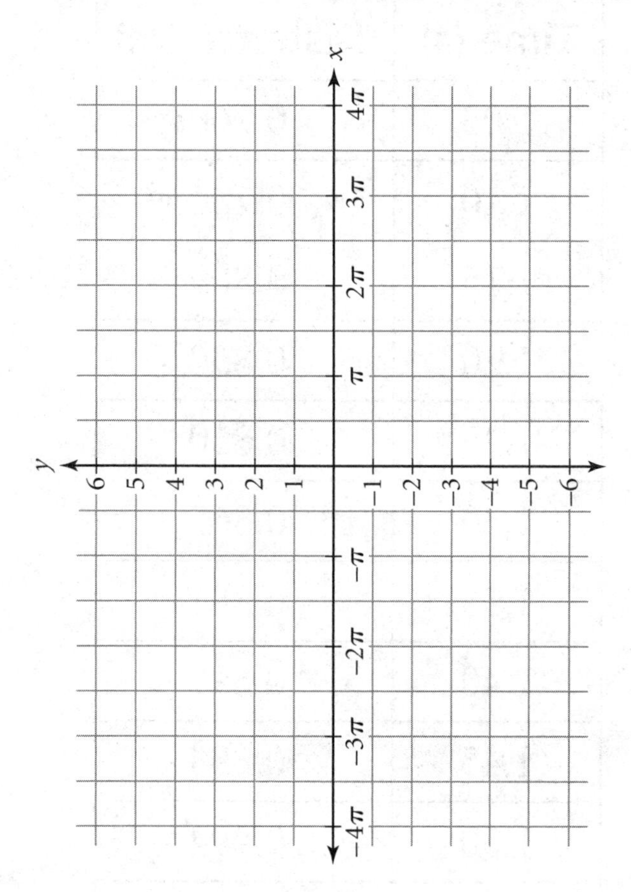

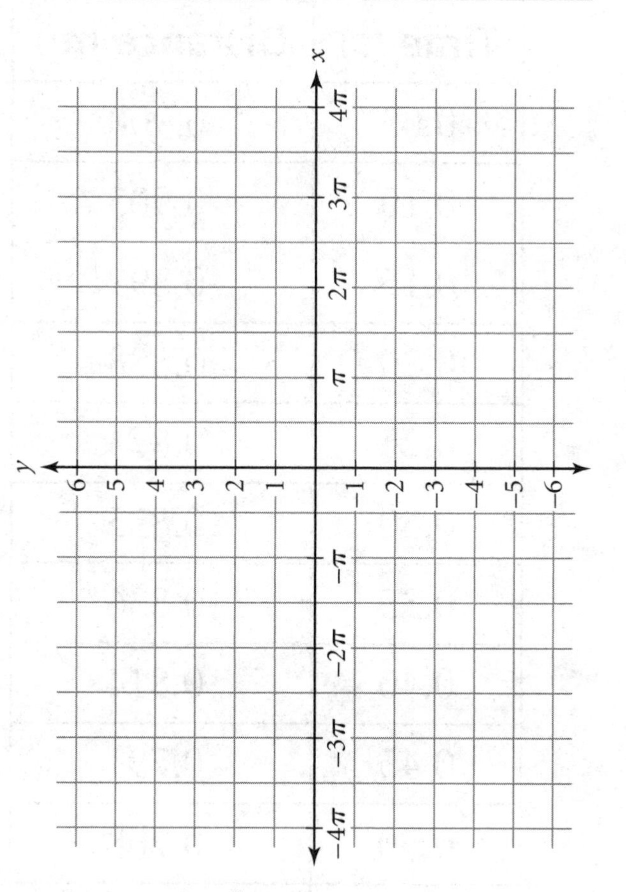

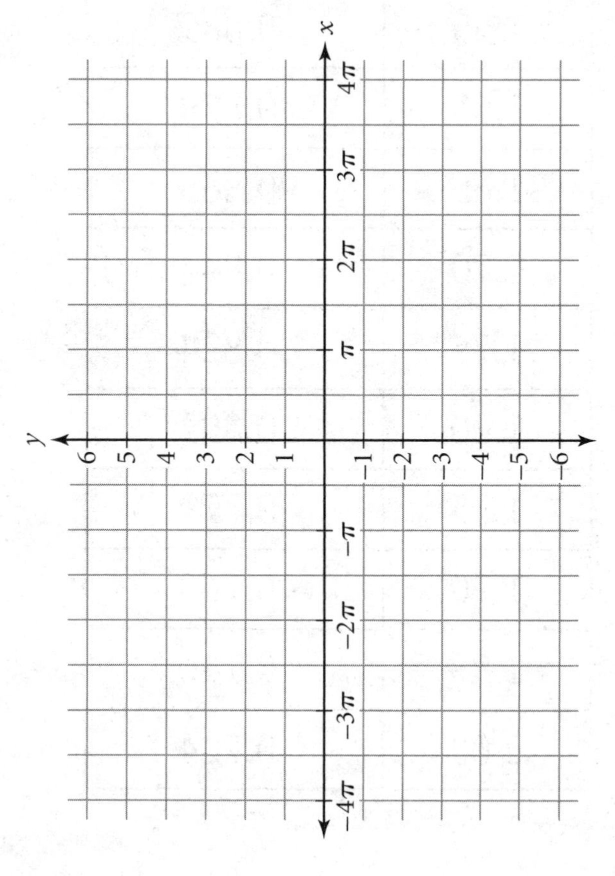

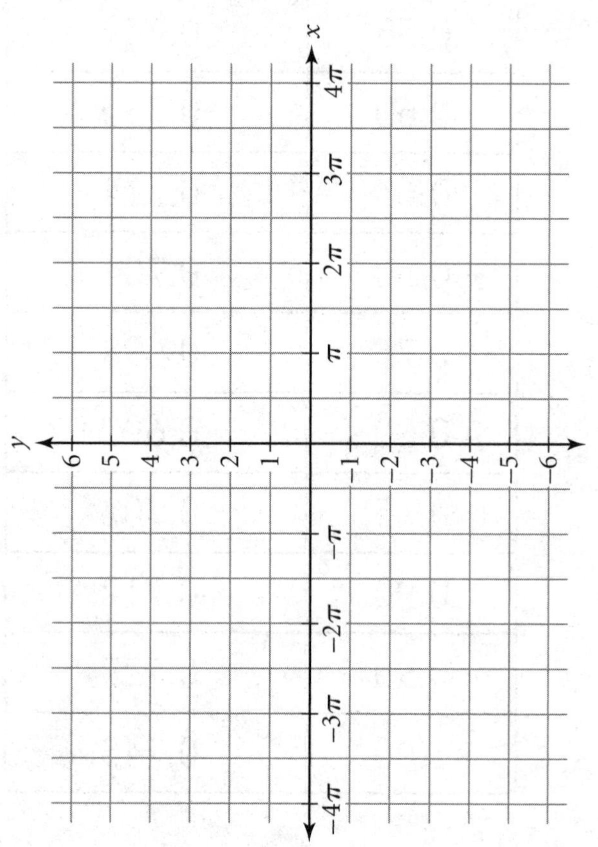

Bouncing Spring Sample Data

Time (s)	Height (m)
0	0.561397
0.05376	0.563182
0.10752	0.578284
0.16128	0.604919
0.21504	0.639792
0.2688	0.673841
0.32256	0.699653
0.37632	0.71805
0.43008	0.724229
0.48384	0.719698
0.5376	0.729446
0.59136	0.701163
0.64512	0.632653
0.69888	0.603409
0.75264	0.576362
0.8064	0.564005
0.86016	0.587071
0.91392	0.597093
0.96768	0.640204
1.02144	0.668624
1.0752	0.705831
1.12896	0.729308
1.18272	0.741253
1.23648	0.738919
1.29024	0.724366
1.344	0.695946
1.39776	0.656954
1.45152	0.620983
1.50528	0.587895
1.55904	0.567712
1.6128	0.5673

Time (s)	Height (m)
1.66656	0.581991
1.72032	0.607665
1.77408	0.630044
1.82784	0.665192
1.8816	0.699241
1.93536	0.720659
1.98912	0.731917
2.04288	0.728622
2.09664	0.715167
2.1504	0.687845
2.20416	0.664505
2.25792	0.631417
2.31168	0.607391
2.36544	0.592975
2.4192	0.586934
2.47296	0.592837
2.52672	0.609038
2.58048	0.632104
2.63424	0.657778
2.688	0.682628
2.74176	0.702261
2.79552	0.713794
2.84928	0.714206
2.90304	0.705419
2.9568	0.690454
3.01056	0.667938
3.06432	0.64144
3.11808	0.618374
3.17184	0.60588
3.2256	0.597505
3.27936	0.597917

Exercise 8

Name ______________________ Period __________ Date ______________

a.

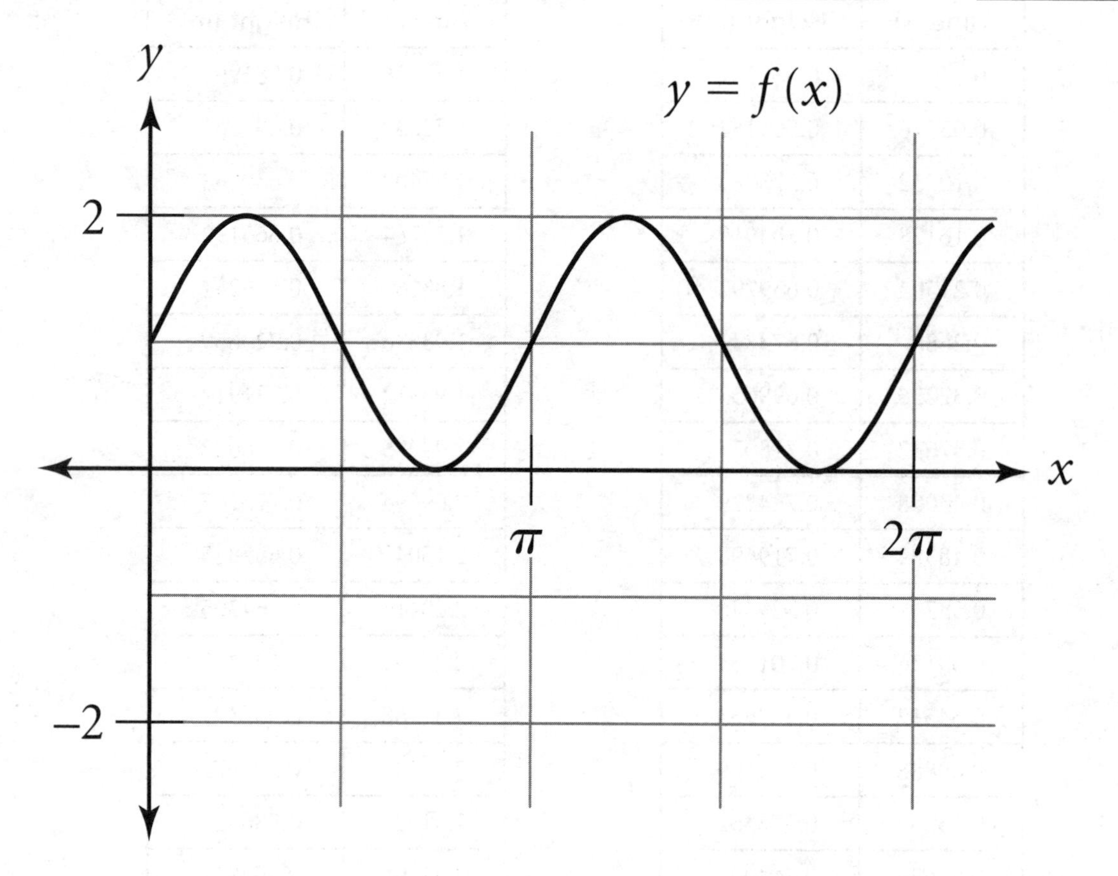

b.

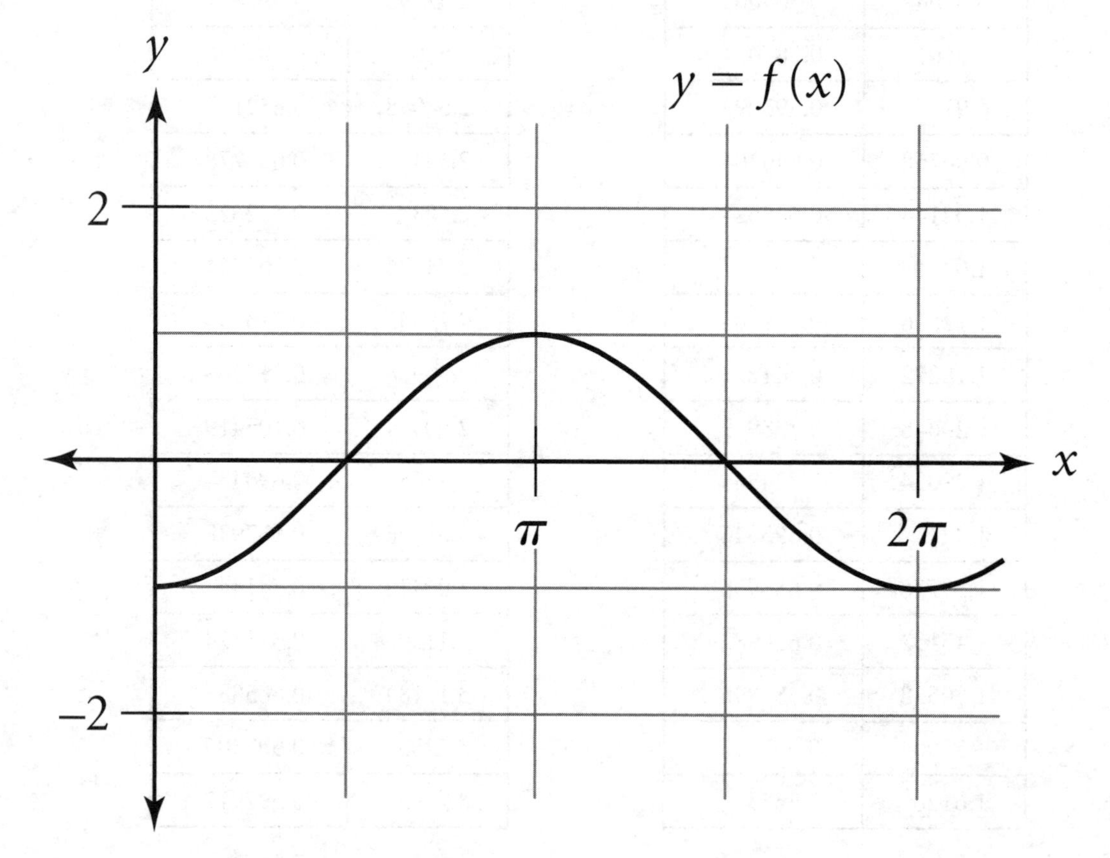

Two Ways to *c*

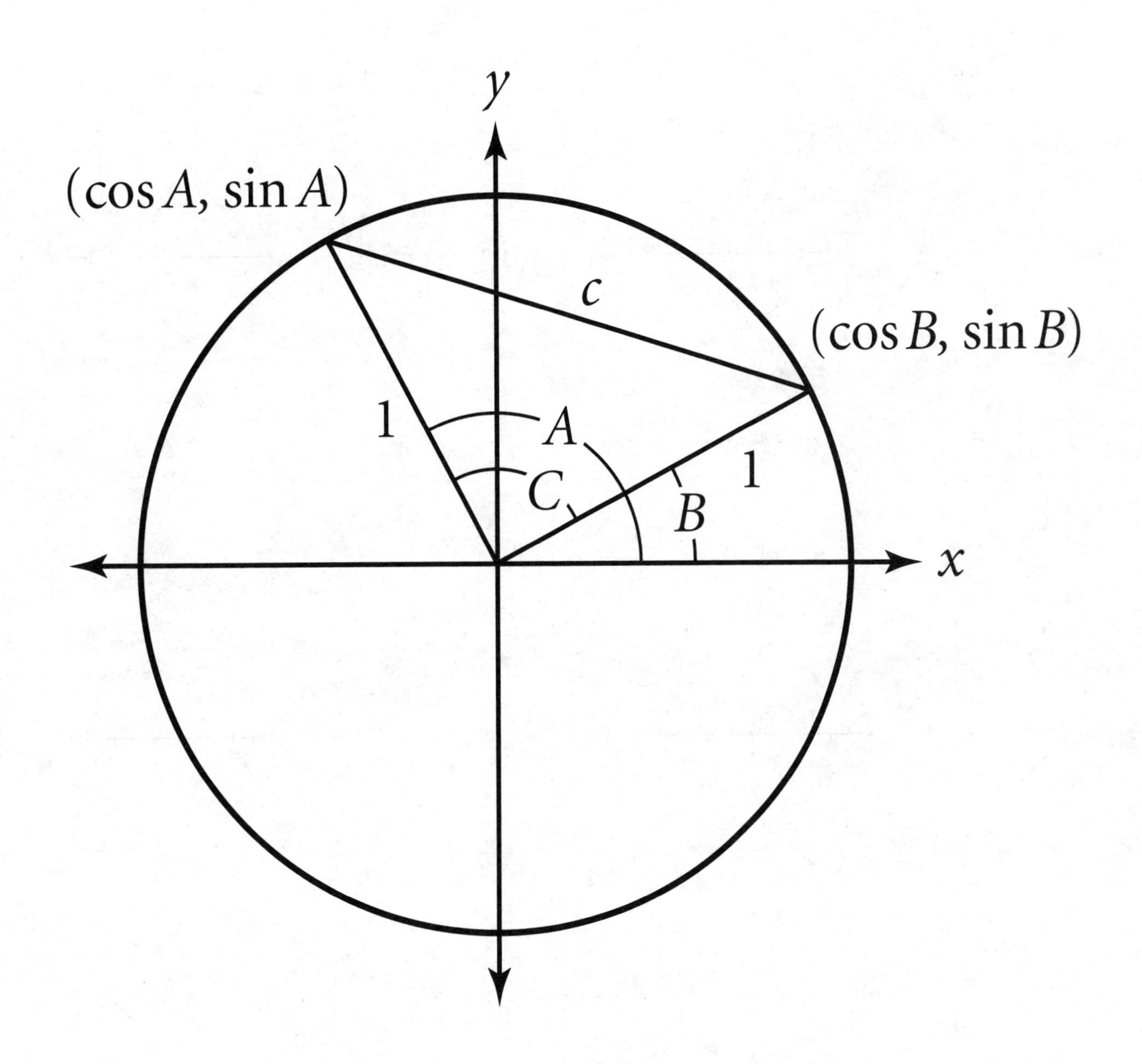

Polar Graph Paper

Name ______________________________ Period ____________ Date ______________

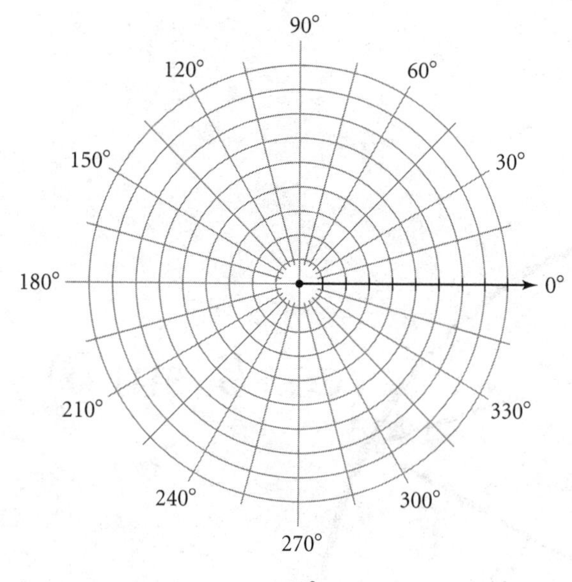

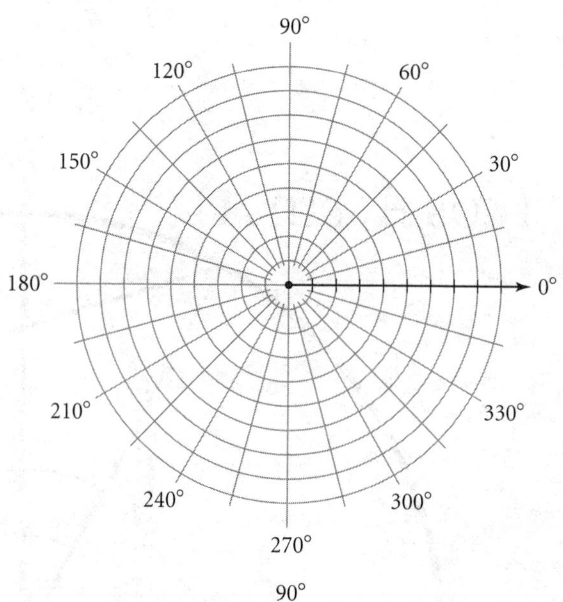

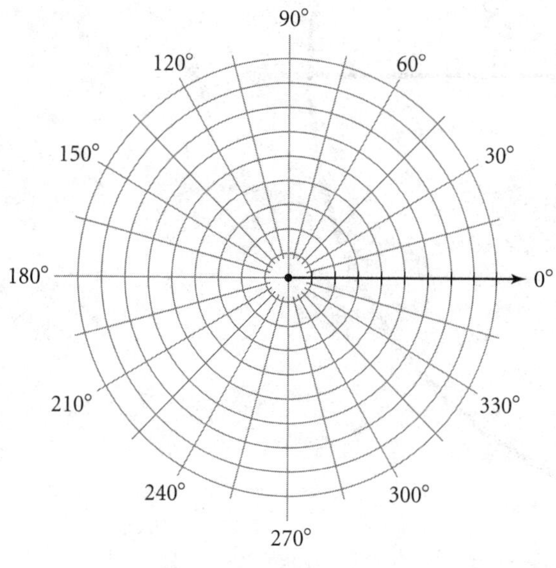

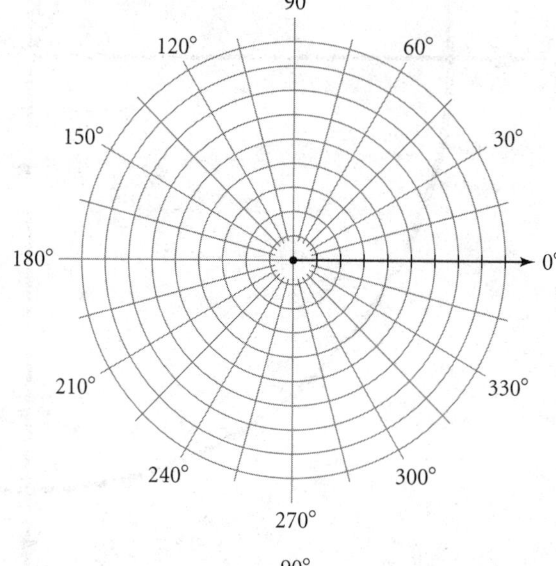

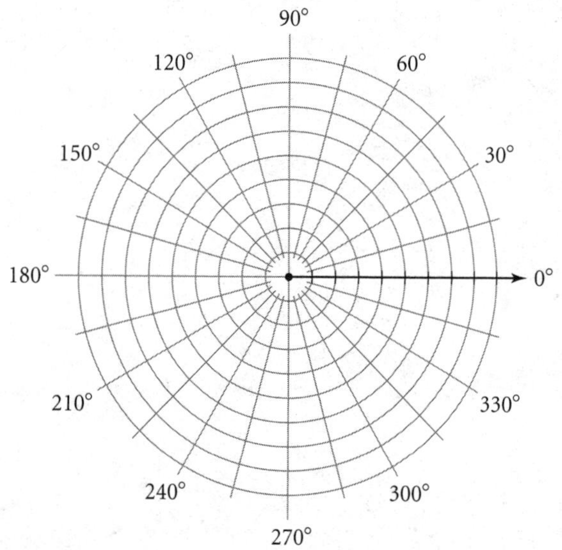

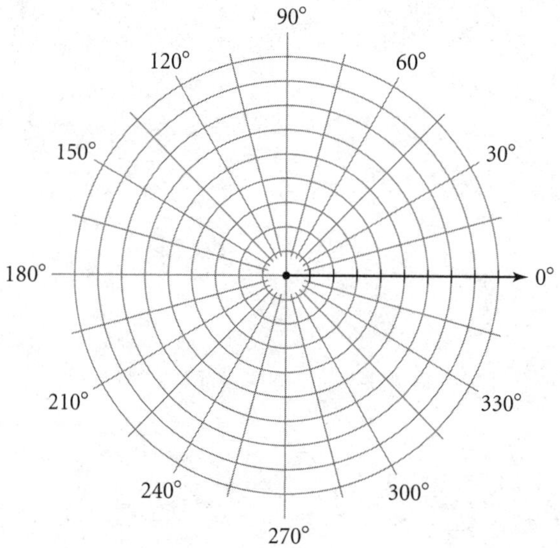

1-Centimeter Grid Paper

Name ______________________________ Period __________ Date ______________

$\frac{1}{4}$-Inch Grid Paper

Name ______________________________ Period __________ Date ______________

$\frac{1}{2}$-Centimeter Grids

Name ______________________ Period __________ Date ______________

Rulers

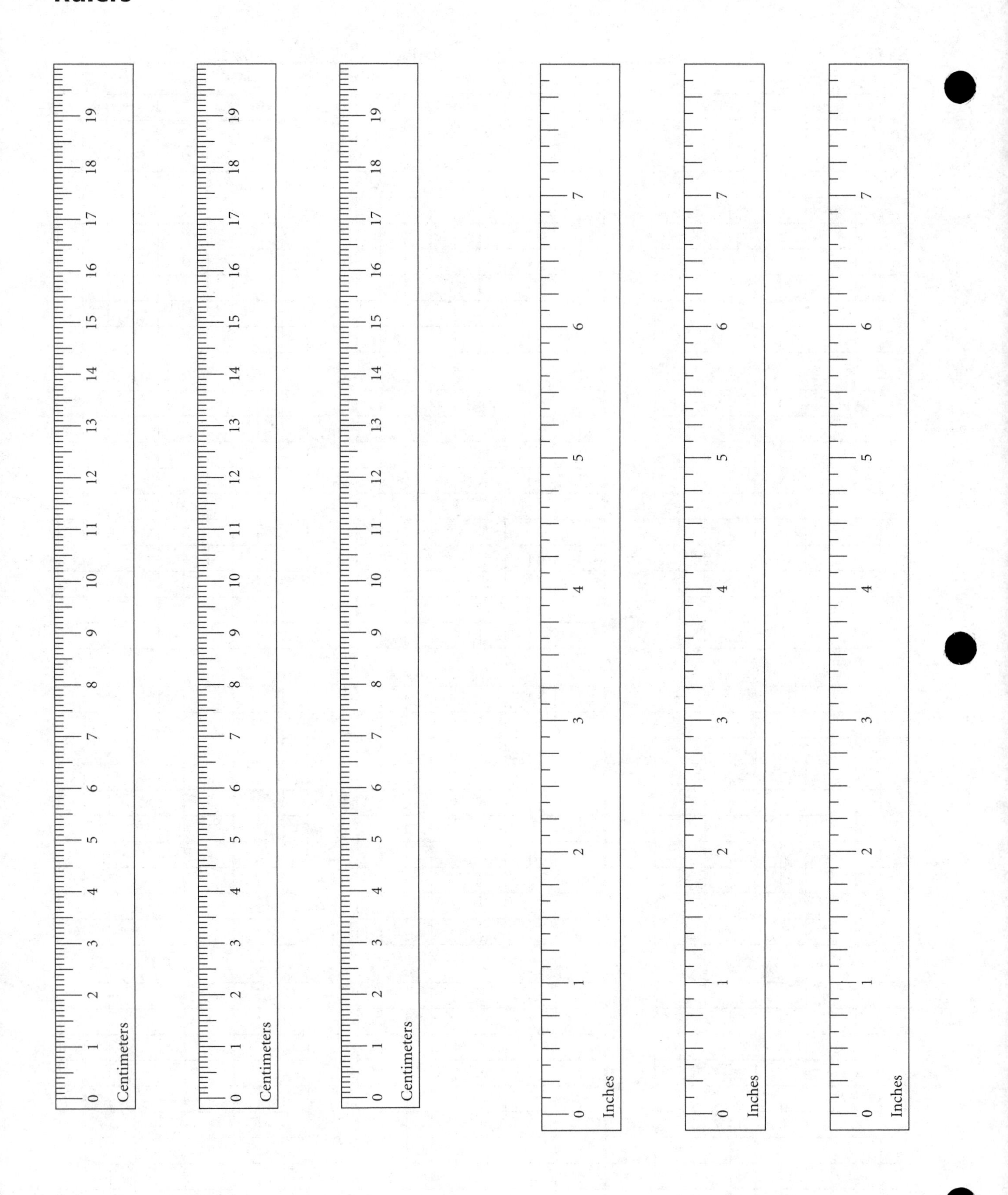

Centimeters
0
1
2
3
4
5
6
7
8
9
10
11
12
13
14
15
16
17
18
19
Inches
0
1
2
3
4
5
6
7

Protractors

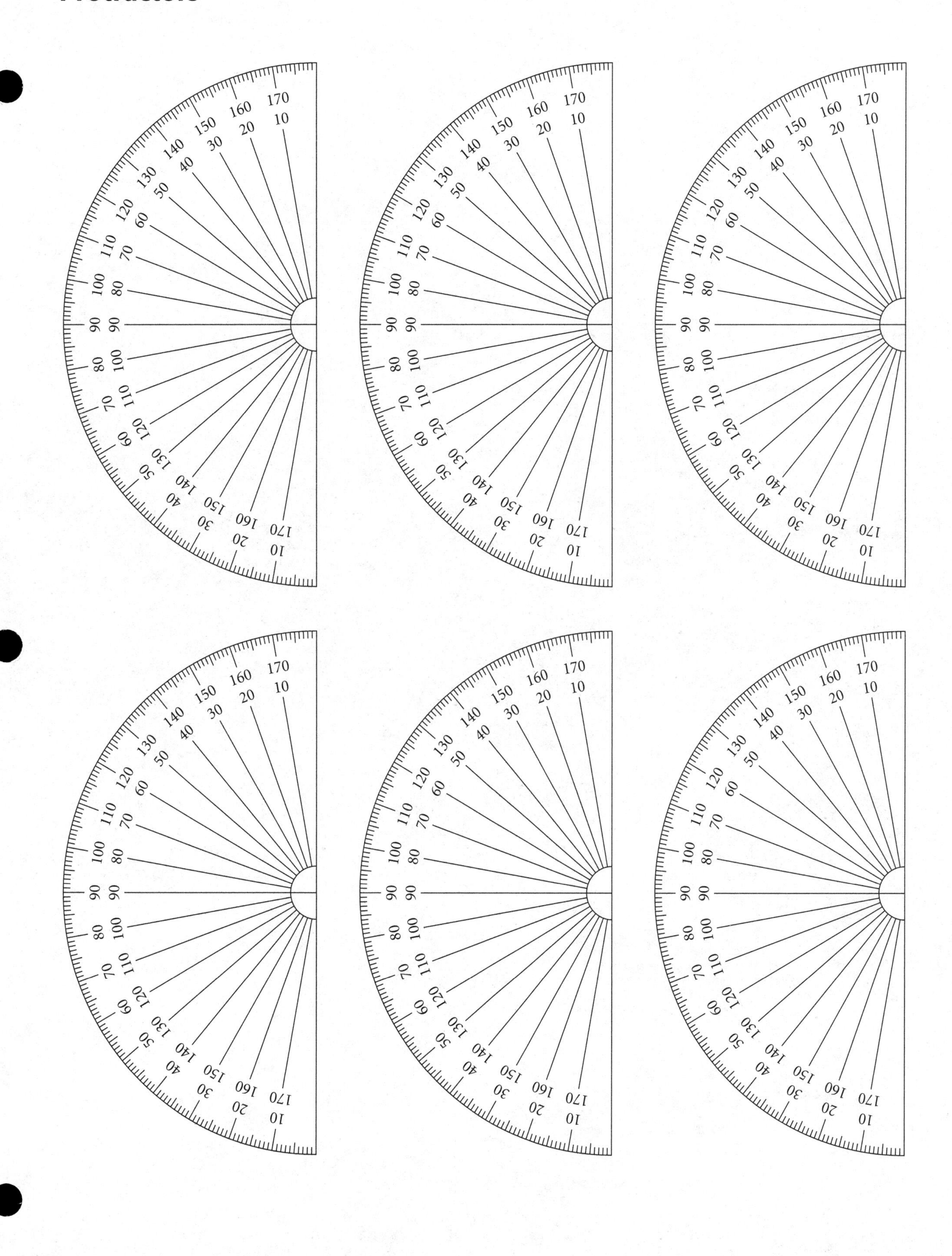

Comment Form

Please take a moment to provide us with feedback about this book. We are eager to read any comments or suggestions you may have. Once you've filled out this form, simply fold it along the dotted lines and drop it in the mail. We'll pay the postage. Thank you!

Your Name ____

School ____

School Address ____

City/State/Zip ____

Phone ______________________________

Book Title ____

Please list any comments you have about this book.

Do you have any suggestions for improving the student or teacher material?

To request a catalog, or place an order, call us toll free at 800-995-MATH, or send a fax to 800-541-2242.
For more information, visit Key's website at www.keypress.com.

Please detach page, fold on lines and tape edge.

NO POSTAGE
NECESSARY
IF MAILED
IN THE
UNITED STATES

BUSINESS REPLY MAIL
FIRST CLASS PERMIT NO. 338 OAKLAND, CA

POSTAGE WILL BE PAID BY ADDRESSEE

KEY CURRICULUM PRESS
1150 65TH STREET
EMERYVILLE CA 94608-9740
ATTN: EDITORIAL

Comment Form

Please take a moment to provide us with feedback about this book. We are eager to read any comments or suggestions you may have. Once you've filled out this form, simply fold it along the dotted lines and drop it in the mail. We'll pay the postage. Thank you!

Your Name ______________________________

School ______________________________

School Address ______________________________

City/State/Zip ______________________________

Phone ____________________

Book Title ______________________________

Please list any comments you have about this book.

Do you have any suggestions for improving the student or teacher material?

Please detach page, fold on lines and tape edge.

NO POSTAGE
NECESSARY
IF MAILED
IN THE
UNITED STATES

POSTAGE WILL BE PAID BY ADDRESSEE

KEY CURRICULUM PRESS
1150 65TH STREET
EMERYVILLE CA 94608-9740
ATTN: EDITORIAL